3D PRINTING TECHNOLOGY FOR WATER TREATMENT APPLICATIONS

Additive Manufacturing Materials and Technologies

3D Printing Technology for Water Treatment Applications

Series Editior

MA QIAN

Edited by

JITENDRA KUMAR PANDEY

University of Petroleum and Energy Studies, Department of Research and Development, Dehradun, India

SUVENDU MANNA

University of Petroleum and Energy Studies, School of Engineering, Dehradun, India

RAVI KUMAR PATEL

University of Petroleum and Energy Studies, Dehradun, India

ELSEVIER

Elsevier
Radarweg 29, PO Box 211, 1000 AE Amsterdam, Netherlands
The Boulevard, Langford Lane, Kidlington, Oxford OX5 1GB, United Kingdom
50 Hampshire Street, 5th Floor, Cambridge, MA 02139, United States

ISBN: 978-0-323-99861-1

For information on all Elsevier publications visit our website at https://www.elsevier.com/books-and-journals

Publisher: Matthew Deans
Acquisitions Editor: Dennis McGonagle
Editorial Project Manager: Isabella C. Silva
Production Project Manager: Prasanna Kalyanaraman
Cover Designer: Christian J. Bilbow

Typeset by TNQ Technologies

Contents

4. An overview of water pollutants in present scenario

Santhosh Naraseeyappa Kuchangi, Manohara Halanur Mruthunjayappa, and Nataraj Sanna Kotrappanavar

5. A brief overview on advances in water treatment process

Amit Kumar, Leena V. Bora, Nikita P. Chokshi, Gajendra Kenil, and Mansuri Zaid

6. 3D-Printed membrane for water treatment

A. Manmadhachary, Roopa Tulasi, and Priyanka Chattoraj

Contributors

Rashi Agrawal School of Engineering, University of Petroleum and Energy Studies, Dehradun, Uttarakhand, India

L. Ananda Shruthi Department of Chemical Engineering, Manipal Institute of Technology, Manipal Academy of Higher Education, Manipal, Karnataka, India

Piyali Basak School of Bioscience and Engineering, Jadavpur University, Kolkata, West Bengal, India

Prakash Bobde University of Petroleum and Energy Studies, Dehradun, Uttarakhand, India

Leena V. Bora School Engineering, Nirma University, Ahmedabad, Gujarat, India

Priyanka Chattoraj Department of Mechatronics, ICFAI Tech, Faculty of Science and Technology, ICFAI Foundation for Higher Education, Hyderabad, India

Chetan Chaudhary School of Engineering, University of Petroleum and Energy Studies, Dehradun, Uttarakhand, India

Nikita P. Chokshi Department of Chemical Engineering, Institute of Technology, Nirma University, Ahmedabad, Gujarat, India

Pratik Das School of Bioscience and Engineering, Jadavpur University, Kolkata, West Bengal, India

Pritam Dey Department of Chemical Engineering, National Institute of Technology Agartala, Agartala, Tripura, India

Sunil Dutt Department of Chemistry, Govt. Post Graduate College Una, Himachal Pradesh, India

Soney C. George Center for Nanoscience and Technology, Amal Jothi College of Engineering, Koovappally, Kerala, India

Shikha Jain Department of Biomedical Engineering, Manipal Institute of Technology, Manipal Academy of Higher Education, Manipal, Karnataka, India

Deepanshu Kashyap School of Engineering, University of Petroleum and Energy Studies, Dehradun, Uttarakhand, India

Gajendra Kenil Department of Chemical Engineering, Institute of Technology, Nirma University, Ahmedabad, Gujarat, India

Santhosh Naraseeyappa Kuchangi Sustainable Materials and Processes Lab, Centre for Nano & Material Sciences, Jain Global Campus, Jain University, Bangalore, Karnataka, India

Amit Kumar Department of Chemical Engineering, Institute of Technology, Nirma University, Ahmedabad, Gujarat, India

Ashwani Kumar Department of Chemistry, Govt. College Kullu, Himachal Pradesh, India

Raj Kumar Department of Pharmaceutical Sciences, University of Michigan, Ann Arbor, MI, United States

Anindita Laha Department of Chemical Engineering, Manipal Institute of Technology, Manipal Academy of Higher Education, Manipal, Karnataka, India

A. Manmadhachary Department of Mechatronics, ICFAI Tech, Faculty of Science and Technology, ICFAI Foundation for Higher Education, Hyderabad, India

Tarun Mateti Department of Chemical Engineering, Manipal Institute of Technology, Manipal Academy of Higher Education, Manipal, Karnataka, India

Manohara Halanur Mruthunjayappa Sustainable Materials and Processes Lab, Centre for Nano & Material Sciences, Jain Global Campus, Jain University, Bangalore, Karnataka, India

Lalit Nagapurkar AIC-Prestige Inspire Foundation, Indore, Madhya Pradesh, India

Sukanchan Palit Department of Chemical Engineering, University of Petroleum and Energy Studies, Dehradun, Uttarakhand, India

Ravi Kumar Patel AIC-Prestige Inspire Foundation, Indore, Madhya Pradesh, India

Sreelakshmi Rajeevan Center for Nanoscience and Technology, Amal Jothi College of Engineering, Koovappally, Kerala, India; APJ Abdul Kalam Technological University, CET Campus, Thiruvananthapuram, Kerala, India

Srimanta Ray Department of Chemical Engineering, National Institute of Technology Agartala, Agartala, Tripura, India

Aditya Sachdeva School of Engineering, University of Petroleum and Energy Studies, Dehradun, Uttarakhand, India

S. Samantaray Department of Mechanical Engineering, ITER, Siksha 'O' Anusandhan (Deemed to be University), Bhubaneswar, Odisha, India

Nataraj Sanna Kotrappanavar Sustainable Materials and Processes Lab, Centre for Nano & Material Sciences, Jain Global Campus, Jain University, Bangalore, Karnataka, India

Darshit Siddhpuria School of Engineering, University of Petroleum and Energy Studies, Dehradun, Uttarakhand, India

Sonika Department of Physics, Rajiv Gandhi University, Itanagar, Arunachal Pradesh, India

Goutam Thakur Department of Biomedical Engineering, Manipal Institute of Technology, Manipal Academy of Higher Education, Manipal, Karnataka, India

Seim Timung School of Engineering, University of Petroleum and Energy Studies, Dehradun, Uttarakhand, India

D.K. Tripathy IIT Kharagpur, Kharagpur, West Bengal, India

S. Tripathy Department of Mechanical Engineering, ITER, Siksha 'O' Anusandhan (Deemed to be University), Bhubaneswar, Odisha, India

Roopa Tulasi Department of Mechanical Engineering, Sreenidhi Institute of Science and Technology, Hyderabad, India

Mansuri Zaid Department of Chemical Engineering, Institute of Technology, Nirma University, Ahmedabad, Gujarat, India

CHAPTER

1

An overview of the advances in the 3D printing technology

Tarun Mateti[2,a], *Shikha Jain*[1,a], *L. Ananda Shruthi*[2,a], *Anindita Laha*[2], and *Goutam Thakur*[1]

[1] Department of Biomedical Engineering, Manipal Institute of Technology, Manipal Academy of Higher Education, Manipal, Karnataka, India; [2] Department of Chemical Engineering, Manipal Institute of Technology, Manipal Academy of Higher Education, Manipal, Karnataka, India

1.1 Introduction

Three-dimensional (3D) Printing may turn a mathematical model into a physical object by gradually depositing material [1]. In 1980, Charles Hull of the United States commercialized the technique of 3D printing by layering structures from computer-aided design (CAD) drawings [2]. It is a groundbreaking technology that has developed into a flexible development platform.

The three-dimensional approach has exploded in popularity recently. The method involves layering thin films such as fluid or powdery polymer, metal, or concrete and then merging the layers. 3D printing promises the possibility of fabricating complex biomedical devices using computer settings and anatomical information from patients and has since produced remarkable gadgets, implants, and frameworks for tissue design, analytics, and medicine conveyance. Access to modest printers fuels the renewed interest in fusing cells with bespoke 3D frameworks for customized regeneration medicine [3]. 3D printing can also create jewelry

[a] Contributed equally.

3D Printing Technology for Water Treatment Applications
https://doi.org/10.1016/B978-0-323-99861-1.00002-3

collections, corneas, rocket engines, and steel bridges, among other things. It provides numerous significant benefits in generating prototypes, including the ease to replicate products, cost-effectiveness, and product security [4].

3D printing creates multiple opportunities for enterprises seeking to increase their outlook. It can currently make graphene-based materials, metals, ceramics, and thermoplastics and has prospects to change businesses and streamline industrial processes.

Utilizing 3D printing will expedite manufacturing while cutting costs. Simultaneously, consumer demand will have a more substantial influence on manufacturing. Clients have a more significant say in the final product and may request its manufacturing to their exact specifications. Meanwhile, 3D printing technology facilities will be customer-centric, with a more adaptable and flexible manufacturing process and enhanced quality control.

3D printing (Fig. 1.1) has many applications in diverse fields like engineering and architecture, health and dentistry, pharmaceuticals, aerospace, and defense. It has pioneered novel materials and methods that are efficient, timely, and cost-friendly. 3D Printing has proven to be the future of modern science and is described in detail in this chapter.

However, many disadvantages exist in utilizing 3D printing in the industrial sector. For instance, its adoption would reduce labor—greatly influencing the economy of countries that depend on low-wage

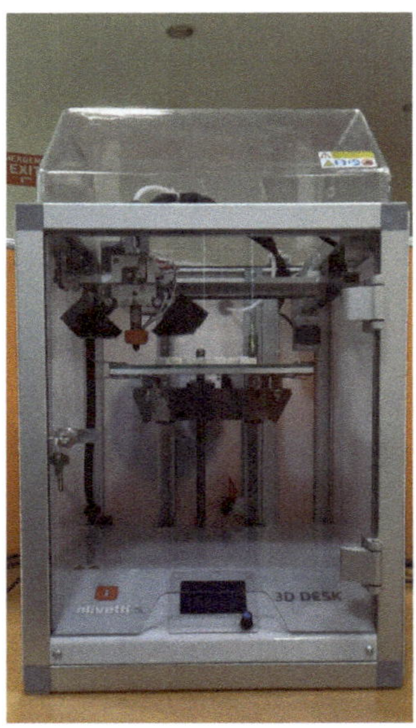

FIGURE 1.1 A 3D printer. *Courtesy: Innovation Center – Manipal Institute of Technology, Manipal Academy of Higher Education.*

employment. Additionally, users may create commodities, including harmful objects like knives, and so, its use must limit only to responsible individuals to prevent lawbreakers. Nevertheless, 3D printing technology has evolved into a flexible and powerful tool for high-tech production [5]. This chapter provides an overview of 3D printing and recent advances in this revolutionary technology.

1.1.1 Chronicle of 3D printing

3D printing was initially introduced during the 1980s and was known as "Rapid Prototyping" because the methods are quick and cost-effective to create prototypes for industrial product development.

1.1.1.1 Foundation of 3D printing—the early 1980s

In 1981, Hideo Kodama of Japan printed layers of material to create a 3D object. Regrettably, Kodama could not get approval for his patent. However, he published a model to produce a printed solid—which served as the foundation for 3D printing [6]. The following decades saw significant advancements in this technology, which has grown into a helpful tool for researchers, manufacturers, designers, engineers, and scientists.

CILAS and the French General Electric Company in France devised a method for creating 3D objects. In any case, the organizations undervalued the idea and quickly abandoned it.

Since Charles Hull built the first 3D printer in 1984 [5], technology has advanced, making machines more functional while their price points decreased, making them more accessible.

In 1986, Charles Hull invented a model for stereolithography. Photopolymers (acrylic-based materials) were used to transform fluid into solid under the influence of intense light. Selective Laser Sintering and Fused Deposition Modeling were also patented during the same time. These models set the groundwork for 3D printing [7].

1.1.1.2 One step ahead—1990s

A few novel 3D printers were devised, including the Solidscap ModelMaker.

New techniques (such as microcasting and splashing materials) enabled the 3D printing of metals. The reception was confined to the creation of high-cost, low-volume items. As a result, it became a well-suited material for developing innovative products in the aviation, automotive, and pharmaceutical industries.

1.1.1.3 The explosion of 3D technology—2000s

While there were several advancements in 3D printing throughout the early 2000s, in 2005, the technology gained mainstream momentum.

Numerous initial patents have expired, and entrepreneurs and investors rushed to seize the opportunity.

Dr. Adrian Bowyer set out to construct a low-cost 3D printer. By 2008, his "Darwin" printer had effectively 3D printed approximately 18% of all its components at the cost of under $650.

When the Fused Deposition Modeling patent was made public in 2009, numerous organizations built various 3D printers, democratizing the technology.

3D printing gained popularity due to the intrigue and critical nature of notions such as printed appendages and kidneys [8].

1.1.1.4 The revolution—2010

As the cost of 3D printers decreased, interest in the invention increased, and they became increasingly prevalent in homes and businesses.

Manufacturers began experimenting with 3D printing. Machine components were rapidly fixed, and supply problems were quickly resolved.

By 2014, 3D printers surpassed the $1 billion mark in sales. With its economic influence, 3D printing has revolutionized the way people work [9].

Individuals were permitted to create new products themselves without the assistance of groups or corporations. This movement fuels "The Maker Revolution"—prioritizing innovation and open-source hardware.

3D printing was expected to grow at a 14.37% compound annual growth rate to about $17.2 billion between 2017 and 2020 [10].

1.1.1.5 COVID-19 and to date

Consistently, the advancement of 3D printing has disrupted the industry. As seen in Fig. 1.2, 3D printing has acquired significant attention for its contributions to education and medicine, along with its extensive commercial use. Various manufacturers and experts stepped forward to combat the COVID-19 pandemic, and the volume of recommendations to offer crucial materials using 3D printing was never expected. Over 1700 professionals volunteered to 3D manufacture veils, respirators, and valves; the possibilities are endless [11].

1.1.2 Computer-aided design (CAD)

Though numerous types of 3D printing techniques exist, most rely on CAD. A manufacturing company must first use a computer to digitally sketch a product before employing a 3D printer to manufacture a prototype.

Computer-aided design, or CAD, creates digital 2D or 3D structures. Two-dimensional CAD models are frequently utilized in architecture and

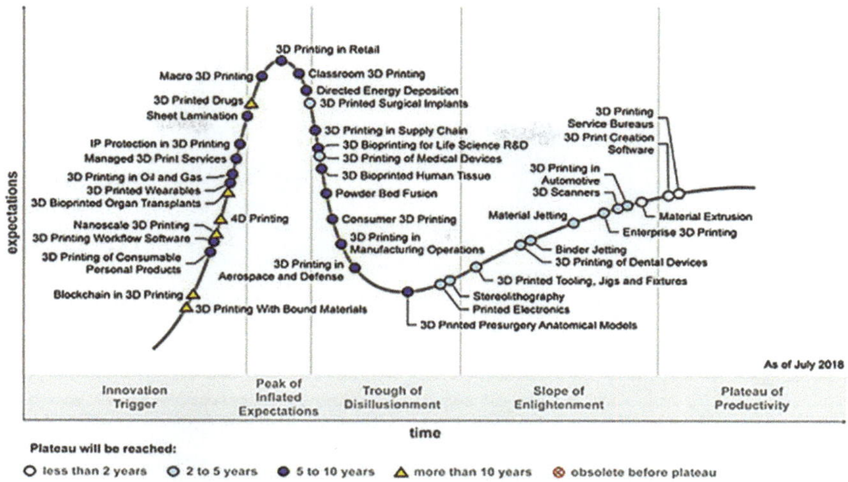

FIGURE 1.2 Hype-cycle of 3D printing [165].

engineering, whereas 3D designs are used in digital innovations such as animation and fabrication operations.

Manufacturing companies employ several forms of CAD to create prototypes and produce final products in proper forms and sizes. While their characteristics differ, they all facilitate the building of computerized models.

3D printers create items using a nozzle coupled to a material filament from which material is ejected onto the print bed. The 3D printer will start working on the next layer after finishing the base layer at the bottom of the print bed.

A CAD file contains guidelines that regulate 3D printers. Data from a CAD file is used to estimate the amount of substance needed and where it is needed. As a result, one must first develop a CAD file to create a suitable dimension model. The CAD file is stored and sent to the 3D printer for processing after the object model is completed. The object will subsequently be built using the directions provided in the CAD file by the 3D printer (Fig. 1.3) [12].

Since being introduced in the 1960s, CAD systems have grown dramatically in product development. CAD systems generate and construct 3D models in engineering, technology, and architecture, setting the foundation for further planning and production [13]. Students may convert their virtual 3D models from CAD to tangible 3D models in this manner. As a result, the student's comprehension of design guidelines appropriate for 3D Printing could be strengthened [14,15].

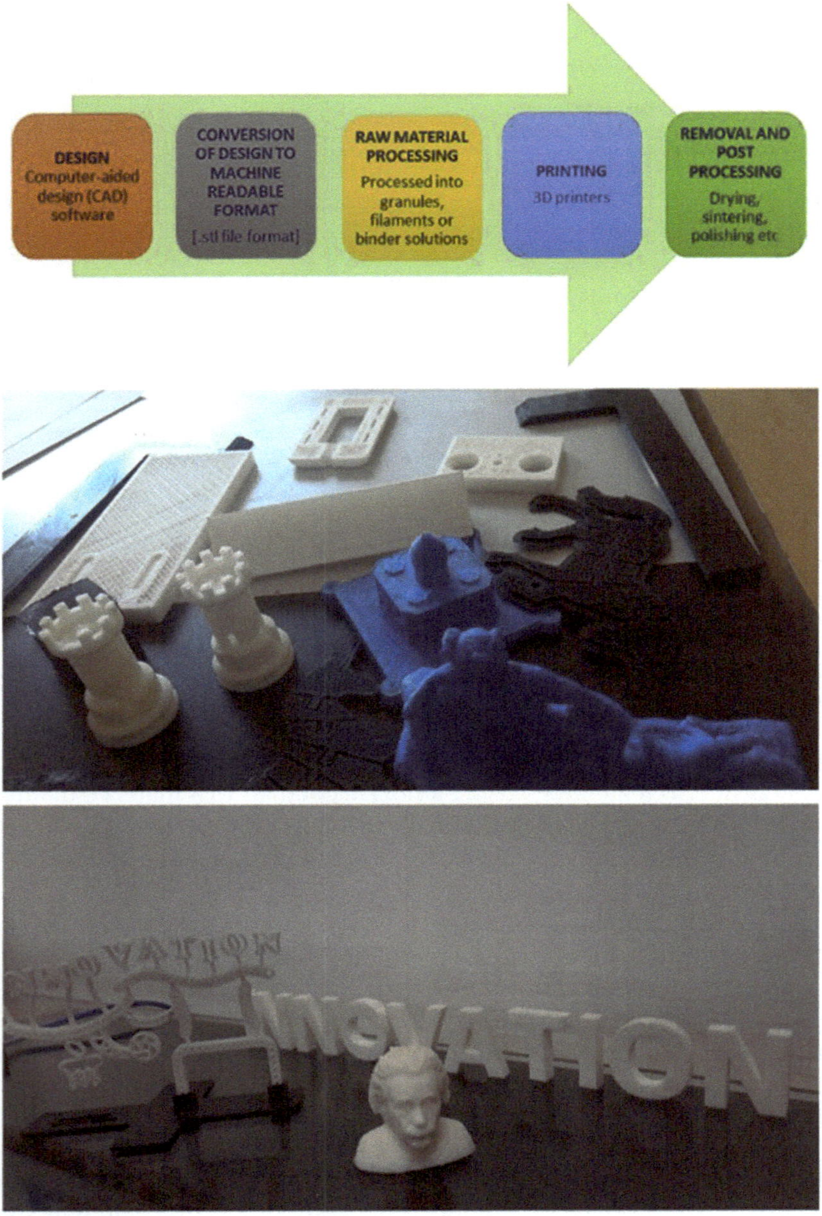

FIGURE 1.3 (Top) overview of 3D printing procedure (middle and bottom) 3D printed objects. *Reproduced with permission A. Mahmood et al., Applications of 3D Printing for the Advancement of Oral Dosage Forms. Elsevier Inc., 2020; Courtesy: Innovation Center – Manipal Institute of Technology, Manipal Academy of Higher Education.*

1.2 Technology classification for 3D printing

Fig. 1.4 describes the types of 3D printing devised so far. A description of the working principle for each type is given below.

1.2.1 Based on material extrusion

Material extrusion began as a quick prototype process but has become a revolutionary production technology for the industrial, medical, automotive, and aerospace industries. This method enables the printing of several materials on plastics, food, and live cells, hence lowering manufacturing costs. Additionally, it can create working components of a product [16].

The material is supplied via a nozzle and placed along a predetermined route to layer-by-layer construct components by heating and extruding. After cooling and solidifying, a tangible item is formed. The earliest example of a material extrusion technique is fused deposition modeling, which utilizes polymer as the principal material [17].

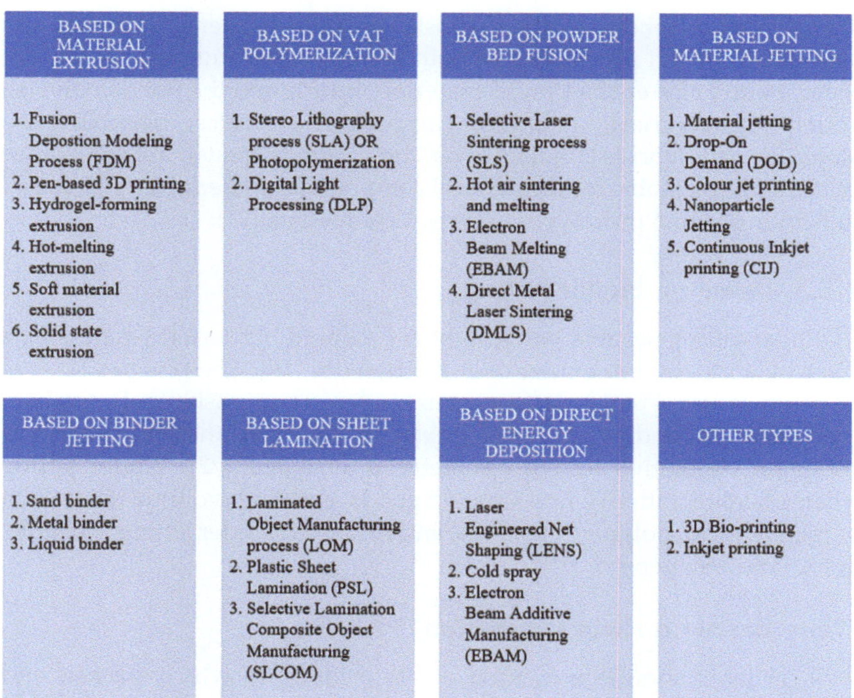

FIGURE 1.4 A brief technology classification for 3D printing.

1.2.2 Based on vat polymerization

Photopolymerization: the curing of photoreactive polymers using UV radiation, light, or laser, is the most often used 3D printing technology. Vat polymerization is a 3D printing technology that hardens a tiny layer of liquid plastic by carefully directing light to a particular location on it. The process continues until the 3D component is created. Stereolithography and digital light processing are two examples of photopolymerization-based 3D printing technology [18].

1.2.3 Based on powder bed fusion

Selective laser sintering, selective heat sintering, and electron beam melting are the different types of powder bed fusion techniques. These approaches melt or fuse powder according to specified dimensions and layer thicknesses using an electron beam or a laser. Materials like ceramics, composites, and metals are used in this process [19].

1.2.4 Based on material jetting

Material jetting is a procedure where the material is cured drop by drop on a build plate. Under ultraviolet (UV) light, a printer distributes droplets of a photosensitive substance that hardens, layering the component [20]. It produces a smooth surface finish and high dimension accuracy and also enables multi-material printing and the use of a diverse variety of materials, including composites, ceramics, polymers, biologicals, and hybrids [16]. However, the products created are sometimes brittle and unsuitable for mechanical components. Material jetting creates full-color product prototypes and medical models [21].

1.2.5 Based on binder jetting

Binder jetting secures particles using a liquid binder and generates a layer by spraying the binder over the powder [18]. The powder layer is distributed over a build platform, and binders are deposited selectively in each layer depending on CAD model information and the process is continued until the desired geometry is obtained [22]. Binder jetting creates casting patterns, raw sintered goods, and high-volume items such as metals, sand, polymers, hybrids, and ceramics. Binder jetting is simple, quick, and inexpensive [18].

1.2.6 Based on sheet lamination

Sheet lamination is a method of 3D printing in which stacked and laminated sheets of bonded material are machined to produce an object [20]. Metal sheets, paper, and plastic are examples of materials utilized in this process. Different sheet lamination methods are utilized for distinct objectives and are intrinsically linked to each procedure. The process is economical, can produce color prints, and facilitate material handling. Furthermore, surplus material may be reused [21].

1.2.7 Based on direct energy deposition

Direct energy deposition is used to repair or augment existing components [16]. Before depositing it with CAD, it melts material using a concentrated energy source (laser or electron beam). This method utilizes metal powders such as titanium, copper, tin, stainless steel, aluminum, nickel, and cobalt [23–25]. Additionally, the method also applies to ceramics, polymers, and metal hybrids.

Direct energy deposition allows for a great degree of control over the grain structure and may generate objects of excellent quality. Laser-engineered net shaping and laser deposition are examples of this technique [16].

The benefits include saving material and progressively replacing traditional techniques of repairing components—complicated and precise ones, such as mending broken turbine blades or propellers [26].

1.2.8 Other types

1.2.8.1 3D bioprinting

Biologic materials—cells, tissues, and molecules—are assembled to accomplish desired functions in bioprinting [27]. The most often used method is extrusion-based bioprinting [28–31], which enables the incorporation of various material kinds—an approach necessitated by the diversity of biomaterials, cell types, and signaling chemicals [32–34].

This technique is based on the precise stacking of biological components and cell development. The most frequently used biological material deposition and patterning methods are inkjet, microextrusion, and laser-assisted printing [35].

The primary benefits of bioprinting are the capacity to mass create tissue-engineered items, great precision in placing various cell types, and the ability to construct tissue with a high cell density [27,36]. Bioprinting is a relatively new developing technology whose primary value is its capacity to construct live organs [27,35,37].

1.2.8.2 Inkjet printing

Inkjet printers are used for both biological and nonbiological applications. It accumulates material droplets deposited by inkjet printing nozzles to print an object [38]. Inkjet printers often employ thermal or piezoelectric heads. Electrically heated heads create pressure pulses that

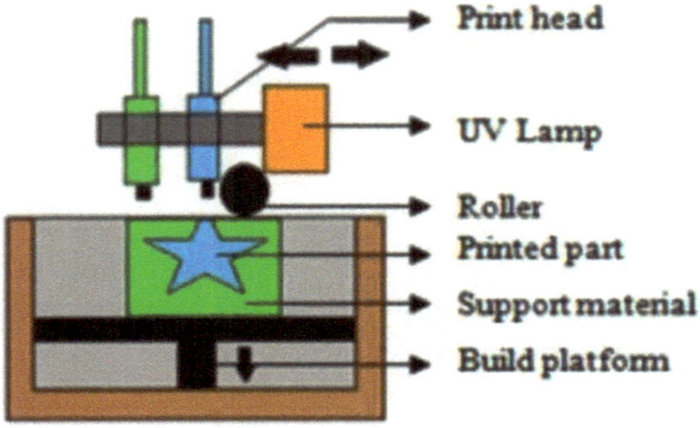

FIGURE 1.5 Polyjet printing [39].

propel droplets out of the nozzle. A crystal located within the head of a printer provides an acoustic pulse that splits the liquid into droplets at equal intervals. When the material is shocked, it rapidly expels droplets from the nozzle due to the resulting pressure [35]. Fig. 1.5 demonstrates polyjet printing (an inkprinting technique) [39], which uses UV light to cure a photo-resin.

1.3 Materials for 3D printing

To create consistent, high-quality products, 3D printing requires materials that fulfill consistent criteria. Material control methods, requirements, and agreements are formed between suppliers, buyers, and end users to accomplish this. 3D printing technology enables the production of completely functioning items from various materials, including ceramics, metals, polymers and their hybrids, composites, and functionally graded materials [17].

1.3.1 Metals

Metals are critical in the aerospace, automotive, and medical industries [40]. They have exceptional physical qualities and may print everything from human organs to aircraft components. Aluminum [41], cobalt-based [42], nickel-based [43], stainless steels [44], and titanium alloys [45,46] are examples of such metals.

Due to their stiffness, robustness, high recovery capacity, and heat resistance, cobalt-based alloys are well-suited for 3D printed dental applications [42]. Additionally, 3D printing technology enables the production of aeronautical components using nickel-based alloys [43].

Nickel-based alloys are suitable for application in hazardous situations because of their strong corrosion and heat resistance (till 1200°C) [40]. Objects using titanium alloys can also be printed, which have unique properties such as ductility, corrosion resistance, oxidation resistance, and low density, and are frequently used in high-temperature environments, such as aerospace components [45,46].

1.3.2 Polymers

3D printers can produce polymer components by depositing successive layers based on material extrusion [47]. Polymers used in 3D printing may be categorized as nanocomposites, particle reinforced composites, fiber-reinforced composites, thermoplastics, thermoresponsive, and thermoplastic composites [39]. Recently, thermoplastics with higher melting points have been employed as 3D printing materials [48]. Due to their lightweight, processing flexibility, and cheap cost, 3D printing polymer materials are frequently employed in the 3D printing industry [49]. Generally, polymers play a significant role in biomaterials and medical devices by providing mechanical support for various orthopedic implants [42].

1.3.3 Ceramics

3D printing may generate ceramics and concrete items without significant pores or fractures. Ceramics are robust, resilient, and resistant to fire. Due to ceramics' fluid condition prior to setting may be put in almost any form, making them ideal for future building [50]. By subjecting ceramics to heat, it is possible to achieve high densities, a highly homogenous microstructure, strong compression strength, and bending [51]. Also, ceramic materials are advantageous for dental and aeronautical applications [52]. Alumina [53], bioactive glasses [51], and zirconia [54] are a few examples.

1.3.4 Composites

Composite materials have revolutionized high-performance sectors with their extraordinary adaptability, lightweight, and tailorable features. Reinforced carbon [55] and glass fiber composites [56] are a few examples of composite materials. Composites are employed in the aviation sector due to their high strength, stiffness, and corrosion resistance [55]. Simultaneously, they have significant potential uses in 3D printing [56], owing to their low cost and outstanding performance [57].

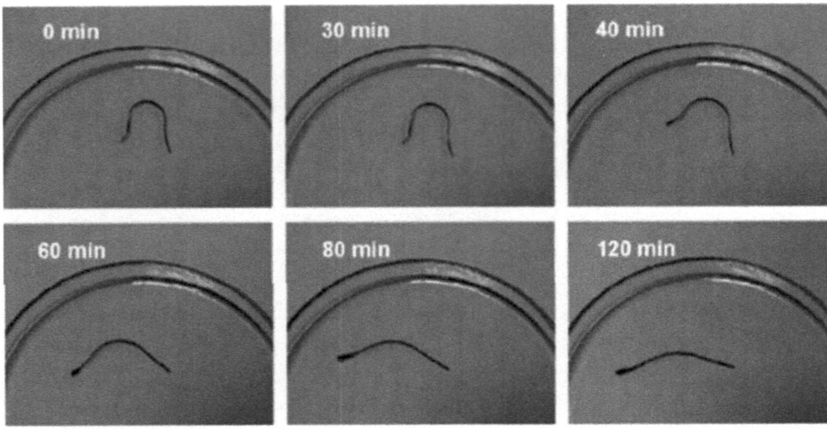

FIGURE 1.6 The shape recovery effect of carbon black/polyurethane. *Reproduced with permission from Q. Meng, J. Hu, A review of shape memory polymer composites and blends, Compos. Part A Appl. Sci. Manuf. 40 (11) (2009) 1661–1672, doi: https://doi.org/10.1016/J.COMPOSITESA.2009.08.011.*

1.3.5 Smart materials

The material complexity enabled by 3D printing has simplified the processing of smart materials: materials that change geometry in response to stimuli such as heat and water [58]. Shape memory polymers (Fig. 1.6) and alloys are two examples of such materials [59,60]. Certain shape memory alloys, such as nickel-titanium [59], have applications ranging from biomedical implants to microelectromechanical systems [50]. They are thermomechanically programmed to achieve complicated geometries in 3D: bends, coils, twists, and folds.

1.3.6 Earth-based materials

Earth-based materials (e.g., fine soil) have attracted considerable attention due to their minimal environmental effect. Despite this, their progress is constrained by the low production rate. Recent advancements have enhanced the design and processing procedures for earth-based materials. Simultaneously, 3D printing has been brought into the construction industry, particularly for cement-based material applications [61].

1.3.7 Novel materials

Novel materials are diverse innovations printed for various purposes. They refer to a sophisticated composite material composed of two or three photopolymers with distinct structures. These create practical prototypes with customizable properties—hardness, texture, and color of the surface [39].

1.3.8 3D printing pen materials

Only a few materials are often utilized in 3D printing pens. Polyvinyl alcohol, polylactic acid, and acrylonitrile butadiene styrene are a few of these materials. The newest 3Doodler pen has a new material palette: polycarbonate and polyamide. Additionally, it works with wood, copper, and bronze composites made of plastics [62].

1.4 Applications of 3D printing technology

1.4.1 Health and dental industry

Three-dimensional printing is used in the medical industry to create drugs, medical devices, bioelectronics, artificial organs, inserts, patches, and orodispersible films (Fig. 1.7) [63–66]. It may be utilized to create polypills containing various medications in distinct compartments, resulting in a range of release characteristics for each drug [67]. Additionally, it is utilized to manufacture tablets. Chewable printlets (to prevent choking), printlets having varying release rates (e.g., burst or sustained), and printlets with distinct features (e.g., high medication loadings) may all be designed to match the specific requirements of each patient. The creation of easily swallowable formulations can significantly boost patient acceptance, especially among the elderly and youngsters. Taste, fragrance, and viscosity all have a role in influencing a formulation's acceptability by youngsters [63].

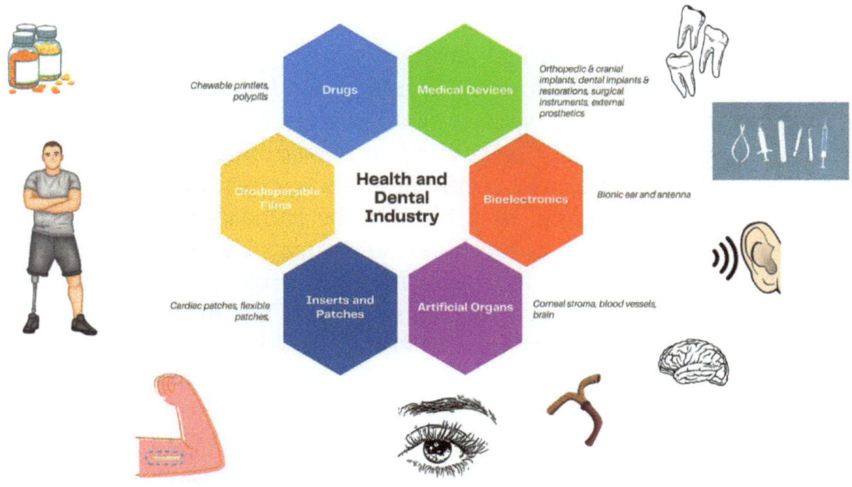

FIGURE 1.7 Applications of 3D printing in the health and dental industry.

The development of orodispersible films can significantly enhance dysphagia patients' acceptance of medicines. This approach might be used to provide medications to nonadherent people since orodispersible films are difficult to remove from the mouth and do not cause gagging intentionally. Veterinary medication distribution is another intriguing use of orodispersible films. Off-label human drugs are often utilized in animal medicine. Introducing orodispersible films for veterinary use set a gold standard for individualizing animal doses [68,69].

The development of bioabsorbable scaffolds dubbed "biopierces"—survives in live tissue while simultaneously expelling antimicrobials to minimize infection throughout the recovery time—is an intriguing medicinal technology. Antimicrobial sensitivity tests showed that biopierces containing the antibiotic successfully discharged the medicine against bacteria. Additionally, it can produce disposable patches that are antibacterial and wound-healing for wound dressing treatments [70,71].

Polydimethylsiloxane, a polymer that is often used in storage systems such as implants, may be produced using ultraviolet light [72]. Additionally, implants and patches have been utilized successfully to treat a variety of malignancies. Flexible patches with high concentrations of the anticancer medicine 5-fluorouracil were created using a polymer combination. The patches were determined to be critical in delivering the medication over 4 weeks and suppressing subcutaneous pancreatic cancer development in mice [66].

Bioprinting was employed to generate vascularized and perfusable cardiac patches anatomically and immunologically compatible with the patient [73]. Since the biolinks used to create these customized patches originated from the same patient, they will not trigger an immune response upon transplantation, obviating the need for immunosuppression. Additionally, cellularized hearts may be printed, implying that bioprinting might be used to replace organs that fail.

In 2018, human corneal stroma analogs were manufactured via bioprinting for the first time. The printed cornea stroma was anatomically identical to a topographically accurate human corneal model. However, further research is essential to determine the cornea's biocompatibility after transplantation and to improve the cornea's capacity to maintain cell proliferation, to name a few. Additionally, a prototype of human skin was created utilizing genuine human skin cells. The bioprinted human skin tissues have been reported to have considerable similarities in terms of cellular structure and content [74,75]. Apart from basic skin prototypes, more complex skin patterns, including an ear, were also produced.

Manufacturing devices capable of multiparameter assessments and replicating or improving the capabilities of live creatures are all components of 3D printing bioelectronics. In 2013, emphasis was laid on developing a bionic ear and antenna. In vitro cartilage tissue was formed around the ear's antenna, allowing for reading signals from electrodes. Both ears were capable of hearing stereo-sound, and the ear was capable of receiving radiofrequency [76].

Drill guides for dental implants; representations for surgery, prosthodontics, and orthodontics; the manufacturing of orthopedic, dental, and craniomaxillofacial implants; and the production of foundations for dental implants are all examples of dental industry applications for 3D printing.

Medical illustration: A medical illustration is an anatomical study model. It enables in-depth investigation of complex, unusual, or unknown anatomy and the planning and practice of surgical methods prior to surgery. 3D printing of a medical illustration leads to speedier, less invasive, and more accurate surgery [77–80].

Drill guides and cutting aids: Drill guides are becoming increasingly prevalent in implant dentistry, and similar technology has been used for total knee replacement in orthopedics. Drill guides are utilized to convey a 3D virtual plan to the operating site, acting as a connection between the digital blueprint and the actual patient [81–84].

Partial dental structures and crown copings: It is possible to build a comprehensive digital replica of the tooth, implant, and dental arch using intraoral optical scanners [85]. Treatment and restorations may be designed by printing crowns and bridges using CAD software [86–88].

Dental models for restorative dentistry: As the usage of intraoral scanners increases, dentists will need to employ 3D Printing to make a replica of the jaw. Patient model data may be kept digitally and printed when needed, significantly lowering storage needs.

Computerized braces: In orthodontics, treatment may be planned, tools created, and wires bent robotically using a digital process that incorporates intraoral or laboratory optical scanning to capture clinical data. The "Invisalign" technology digitally realigns a patient's teeth to generate a set of 3D printed models to fabricate "aligners," which progressively correct the teeth over months/years [84]. Bracket-bonding splints—produced in strong and flexible materials—enable precise bracket insertion using CAD [89].

Dental Implants: Implants with a rough or porous surface have been manufactured using 3D printing [83,90]. Through 3D modeling, complex geometries, including a bonelike structure, may be generated [91].

Instrument development and product engineering: 3D printing enables rapid prototyping of instruments, allowing creative individuals to bring their ideas to life in a short period.

1.4.2 Pharmaceutical industry

The pharmaceutical industry is an early adopter of 3D printing—one with enormous growth potential—owing to the customization capabilities and the ability to improve people's lives [92].

While printing tissues and live organs may seem far-fetched, experts believe that patients using algorithms may ultimately be able to print their prescriptions at home using a 3D printer. The technology can produce a broader variety of pharmaceuticals, altering economies of scale and making any drug cost-effective [92].

When treating patients with varying demands, interindividual heterogeneity is a common occurrence. Since the past decade, dosage modification has relied on empirical approaches, which has resulted in decreased effectiveness and an increased risk of unwanted consequences. By adopting 3D printing technology to print drugs, efficiency, precise dose control, excellent repeatability, novel drug delivery devices, and 3D printed polypills with complicated drug-release patterns may be achieved [93,94].

1.4.3 Engineering

The construction business is well-positioned to gain from 3D printing. It can increase safety, save labor and time, and develop personalization. Perhaps the most compelling reason to use 3D printing in construction is the decrease in labor needs, resulting in cost and time savings and a higher degree of site safety, especially in difficult or hazardous situations. Additionally, automated construction might help reduce expensive mistakes and flaws while increasing safety. Additionally, 3D printing eliminates design constraints [95] and offers a structural advantage and aesthetic.

Most research in chemical engineering has concentrated on optimizing mass transfer through shape optimization. Diverse ways for incorporating printable and nonprintable functional components expand the range of applications for digitally manufactured reactors. Electrochemical processes, for example, have been carried out using electrodes that were directly printed as part of the reactor design or through exchangeable electrodes connected via a printed threaded port [96,97]. Similarly, it was proved that batch and flow reactors might be integrated [96,98,99].

3D printing offers extraordinary promise for tissue engineering because it can provide a quick and robust method for in vitro assembly of functional tissue [100–104]. A functioning tissue needs a unique design that offers structural and mechanical support, an adequate supply of nutrients, the required cell types, and the capacity to remodel actively once transplanted [105]. Through biomaterials, printing processes, and cell transport technologies, 3D printing presents an efficient approach for assembling these essential components. Initially, complex scaffolds were printed, followed by a cell seeding method [105,106], but recent efforts strive to simplify processes and concurrently supply structure and cells through scaffold-based or scaffold-free designs [101,102,107–109].

In recent years, scientists and engineers in biomedical engineering have printed organs [35], implants and prostheses [110–112], drugs [113], and models for educational [114]. The technique of fabricating the organ or body part is identical to that of fabricating a plastic or metal component, except that the raw materials are biological cells generated in a laboratory. Customizing the cells for a specific patient may ensure that the patient's body will not reject the organ.

3D printers may also be used to create surgical equipment [115–118]. 3D printing provides great flexibility in creating biomedical equipment since it eliminates the geometric constraints associated with conventional manufacturing processes.

Since biomedical devices and implants must comply with the patient's anatomy, it is challenging to create them using conventional manufacturing processes. Additionally, these devices are intended to be patient-specific [119]. 3D printing eliminates the need for extra production phases such as line installation and mold creation, allowing for quick component fabrication. This makes 3D printing an excellent approach for producing biomedical equipment. Due to its great geometric precision and resolution, 3D printing has been a commonly accepted method in the biomedical field. Additionally, the capability to print sophisticated body implants using magnetic resonance imaging (MRI) [120] and computed tomography (CT) [121] data expanded this technology's utility. 3D bioprinting is another application that allows for the construction of sophisticated living tissues and artificial organs [122,123]. These printers are often made of hydrogel, silicon, or protein solutions. The discipline's primary objective is to rapidly develop functioning and transplantable human organs [124].

1.4.4 Architecture

The notion of constructing a home or structure utilizing robots and minimum human work is one of the most cost-effective, environmentally friendly, and cost-effective methods of construction available today.

Contour crafting is a method for rapidly fabricating large-scale structures in a layer-by-layer approach. The procedure reduces the cost of creating a structure by reducing the number of man-hours required to make it. It also enhances the speed with which the home can be sold, cutting the financing cost by enabling the builder to profit more rapidly [125].

Additionally, the approach is significantly safer for construction workers than conventional building processes, which reduces insurance costs for construction organizations.

1.4.4.1 Temporary housing

Contour crafting can construct a 2000-square-foot home equipped with full electrical and plumbing services in less than 24 h using waste material on-site. It enables millions of individuals who have been rendered homeless worldwide due to natural disasters, wars, and riots to get shelter while receiving other humanitarian relief like food, water, medication, and clothes. In these instances, the system that requires the least amount of work is optimal [125].

1.4.4.2 Housing for the poor

Housing is an issue in emerging nations because they lack the economic riches necessary to create cities at the pace of our population expansion. Another issue is the growth of slum homes inside cities. If contour crafting is utilized in regions where slum development is common, it improves the quality of life in such areas and slows the rise of slums. Contour crafting addresses many of the issues associated with the Indian housing crisis while also providing women in India with an opportunity to join in the building sector, defying societal standards yet portraying India well in the world arena [125].

However, polymer-based 3D printing materials lack the fire resistance needed by many locations and occupancies due to their low tensile strength, anisotropy, and temperature sensitivity. Mesh Molds are advantageous because they are made of continuous fiber polymer, which has a higher mechanical strength. Combining polymer composites with cementitious elements may comply with building code standards. However, the approaches for mesh model building should be improved [126].

Cementitious-based materials provide structural benefits compared to polymer composites. Cementitious materials have high compressive strength but a low tension strength, necessitating reinforcement. When short and continuous fibers are introduced to cementitious materials, structural strength may be increased. However, there are issues regarding concrete's flexural strength, which would need further reinforcement. The extra reinforcement is not automatic and must be installed manually [126].

3D printing technology provides novel approaches to design and construction. The use of 3D printing in architecture is contingent upon the printer's precision, the availability of a diverse range of materials, and manufacturing costs.

1.4.5 Irrigation

3D printing might be used in precision agricultural applications. Precision agriculture is a concept that emphasizes the use of technology that enables systems to be customized for increased efficiency while still assuring a minimal environmental effect [124,127–131].

Precision irrigation is an intriguing topic of precision agriculture. In this sector, technologies are used to manage the amount of water that must be delivered to the plants in order for them to operate appropriately in the given environmental conditions and plant type, hence avoiding water waste [124,127–133]. Another example of precision agriculture is the widespread use of anti-UV polymer coatings in the horticulture industry. Researchers have been examining its effects on plants since UV-B radiation has been identified as a possible hazard to living beings by altering plant shape, biochemistry, and development. For plants, this form of radiation results in growth retardation, visible injuries, and oxidative damage [134–138].

1.4.6 Education

A 3D printer may be used in a range of educational fields. Students of history, for example, may print historical artifacts for analysis, while students of web design can create printable 3D models of their inventions. Students of geography may print maps of geomorphology, socioeconomics, and census. Students of chemistry may make 3D representations of particles. Biology students may synthesize cells, viruses, organs, and other biological items, and students in math may print out 3D representations of the problems they are working on [139–144].

The following are the key educational benefits of 3D printing:

- **Stimulates curiosity and interest.** The student may be involved in the research and design phase through their manufacturing.
- **Creates an atmosphere of enthusiasm and active engagement.** It prevents students from becoming passive consumers of information. The learner may now have an active role in developing products related to the lecture material, rather than being a passive consumer of knowledge.
- **Provides new learning possibilities.** The degree of learning is increased by expanding the amount of information acquired in a context via 3D objects.
- **Promotes the development of problem-solving skills.** To print items correctly, students must learn to recognize and address real-world issues, improve their manual abilities, and persist in the face of adversity. Additionally, by making mistakes and repeating

procedures until they succeed, students will gain confidence in executing more challenging ideas and efforts, expanding their expertise.

- **Assistive technology.** Three-dimensional printing may also benefit kids with specific needs, such as the visually impaired or blind. For example, the Tactile Picture Books initiative designed books for visually impaired children. The project provides crucial assistance to disabled youngsters and allows students to experiment with 3D printing [139–145].

Revolutionary solids, two-variable functions, and 3D representations of the graph of a real function of one variable are only a few of the integral calculus applications of 3D printing. Graph printer is one of the numerous tools that enable users to create 3D representations of function graphs by inputting the function term and the interval. Additional characteristics, such as the addition of specified coordinate axes or the whole coordinate system, may be adjusted [146].

According to studies, students seldom use and comprehend concepts such as continuity and differentiability. By touching the pattern with their fingertips, students may "feel the graph's continuity as dynamic continuity." Using an extended hand to move around the model may highlight the changing slope. The graphs covering regions show local linearity. These efforts may serve as a conceptual foundation for future formalism. They may be conducted using 3D printed prototypes, allowing for the qualitative development of concepts such as continuity and differentiability [147,148].

3D printing prototypes have been developed as a research and scientific teaching tool, particularly in medical science education. Two foam NaCl molds were created using CAD software. Image processing methods were used to construct a 3D representation of a thoracic vertebra. A human vertebra replica was employed to examine the current condition of the human bone architecture. The prototype outputs are compared to the model to ensure that they are equivalent, and according to research, they have an 83% likeness. It may be deduced that all prototypes constructed have high accuracy and are suitable for material study. Following that, the lifelike impression created by the bone mock-up may be employed for further reasons, such as implant placement or surgical planning [149].

The traditional apprenticeship model of surgeon education has downsides, including greater patient risk. 3D printed simulators overcome this barrier by enabling trainees to practice standard surgical methods and expose qualified physicians to advanced skills before the actual surgery.

Combining 3D printing with high-fidelity standards or haptic feedback simulates a range of intraoperative scenarios. As a consequence of these calculations, feedback data on the face, structure, and construct validity of different models have been created for use in various studies [150–155]. Additionally, validation experiments have shown that the resultant virtual applicability in medical education is reproducible and may be utilized in place of cadaveric disassembly [156,157].

The adaptability of 3D printed surgical models is one strategy for bridging the global education barrier. Developing new materials or reinventing the use of existing materials with more functional integrity should be pursued in higher-income countries since advancements in surgical technology are conceivable.

1.4.7 Food industry

Fig. 1.8 depicts the food products—chocolates, cakes, jelly, pizza, cheese, sugar, peanut butter, meat, protein bars, syrups, fries, puddings and rolls, among much other delicacies—that can be printed using suitable methodologies. These almost retain the taste and feel of the original products while also providing a phenomenal opportunity to tune their nutrition profile. Rather than focusing product production processes in a single phase, printing food may create food with novel textures and improved nutritional value. It is performed by symbiotically combining the key food constituents, considering their inherent properties and binding processes during layer deposition.

Printing inks may be attractive in the food packaging business, where an ongoing demand for safer, smoother, more affordable inks, coatings, and overprint varnishes. The technology can be used to cover plastic containers to preserve the taste and meal. Researchers demonstrated a significant use of 3D printing methods in the food sector by developing the Fab@Home, an open-source 3D printer using a liquid component [158,159].

Three interdependent factors are recognized as critical for rational 3D printing method selection in food design: (1) printability, (2) application, and (3) postprocessing feasibility (Fig. 1.9).

A 3D printer (Fab@home) printed cake icing and sugar biscuits. The dough was printed by adjusting the concentrations of butter, yolk, and sugar. Transglutaminase and bacon fat were used for printing meat purees [158,160].

TNO and Barilla have created additive-free 3D printed pasta. Natural Machines has developed the Foodini Food Printer, which creates meals using fresh food components extruded from a food extruder. 3D printing was also used to print chocolates at temperatures ranging from around 28 to 40°C [161,162].

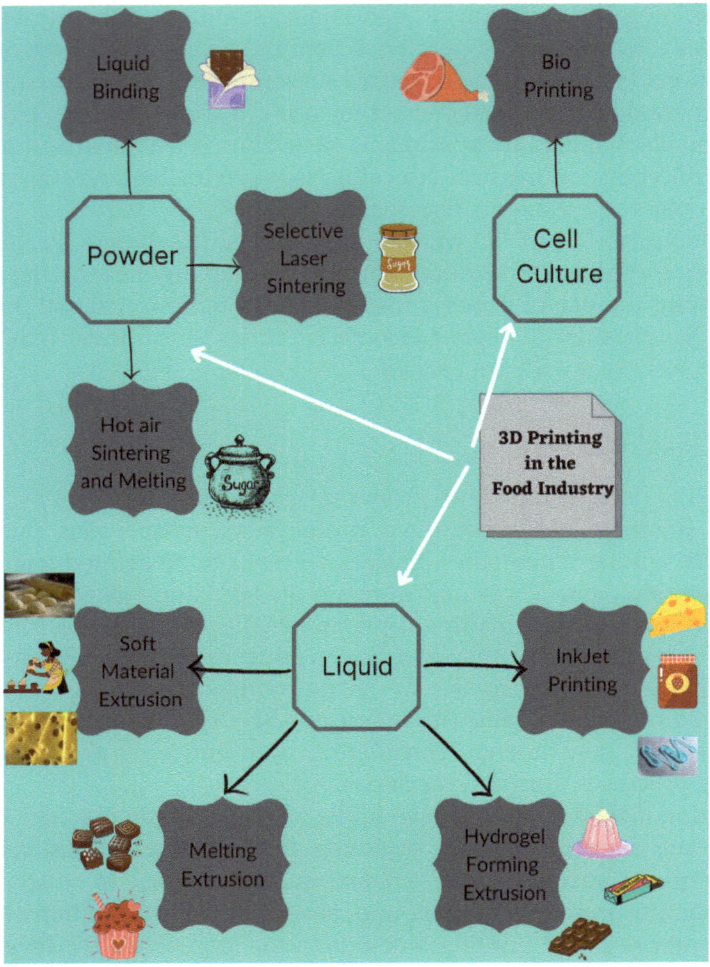

FIGURE 1.8 3D printing in the food industry.

1.4.8 Aerospace and defense

The aerospace and defense industries were early users of 3D printing, using the technology for product development and prototype as early as 1989. Now, three decades later, the aerospace and defense sector accounts for 16.8% of the $10.4 billion 3D printing market and significantly contributes to the industry's continuous research efforts.

Due to the essence of aircraft development, research is intensely rigorous, and maintaining standards is paramount. Numerous essential applications for the aerospace industry have been established via process and material development, and some noncritical components are now flying aboard airplanes.

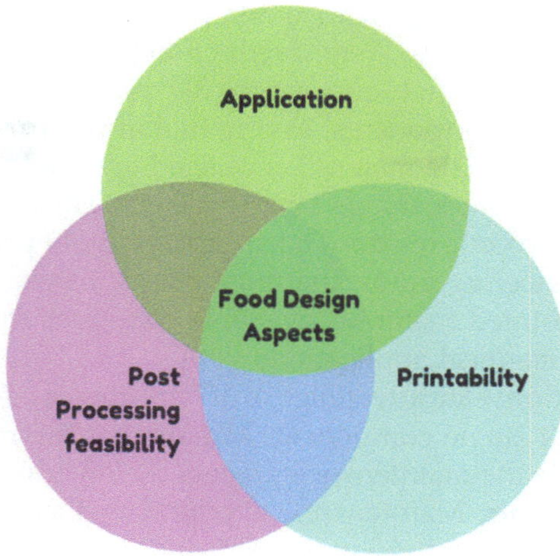

FIGURE 1.9 Food design aspects of 3D printing.

The growth of 3D printing in the aerospace and defense industries is being primarily pushed by industry leaders such as Boeing and Airbus. These and other businesses recognized the value proposition that 3D printing provides—prototypes that function, tooling, and lightweight components [163].

3D printing is not restricted to prototypes in aerospace. Additionally, genuine, working components are 3D printed and integrated into airplanes. Air ducts, wall panels, and even structural metal components are all examples of things that may be made with 3D printing.

The automobile industry was another early user of 3D printing technology. Numerous automobile firms have followed a similar path to aeronautical firms. They use prototype technologies but are also creating and changing their production processes to embrace the advantages of enhanced materials and outcomes for automobile components! Numerous car manufacturers are already investigating the possibility of 3D printing to handle after-sales operations, such as manufacturing spare/replacement components on-demand rather than maintaining a large inventory.

The following are the advantages of 3D printing for aerospace and defense:

Low-volume manufacturing: 3D printing is perfect for aerospace and defense sectors, where extremely complicated components are manufactured in small quantities. Without investing in costly tooling equipment, complicated shapes may be manufactured with the technique. This provides aerospace industry suppliers with a cost-effective method of producing small components.

Weight reduction: Along with other parameters, weight is critical when designing an airplane. Weight reduction may substantially impact an aircraft's emissions, fuel, and cargo. 3D printing makes simple to complex lightweight components, saving fuel in return.

Material efficiency: The material is utilized sparingly because the 3D printing technique creates things layer by layer. Thus, it generates less waste than other procedures do. The spectrum of 3D printing materials ranges from engineering-grade thermoplastics to metal powders. The spectrum of 3D printed materials available continues to grow, enabling sophisticated aeronautical applications.

Part consolidation: One advantage of 3D printing is combining several components into a single component. Minimizing the required components may simplify the maintenance process by lowering assembly time.

Upkeep and repair: Maintenance is critical in the industry, with an aircraft's typical life of up to 30 years. Aerospace and military equipment are often repaired using metal 3D printing [163]. Metal printing is becoming more prevalent in the rocket manufacturing process. Engineers can reinvent the design of rocket components and build them more quickly due to the technology.

In the defense sector, 3D printing can revolutionize the components of military equipment. Defense uses equipment from tiny surveillance drones to jet engines and submarine components.

Electronics 3D printing is a relatively new but rapidly rising area of interest for defense contractors. Engineers can now design and manufacture prototypes of complicated circuit boards and antennas in-house using the technique.

The aerospace and defense industries account for a significant percentage of the 3D printing market. The reasons for this are straightforward: 3D printing has immense potential, from increasing aircraft performance to facilitating more rapid spare component manufacture [163].

1.5 Limitations of 3D printing

3D printing needs to overcome specific obstacles to transition into large-scale manufacturing units. Among them are the following:

Strength is less than traditional manufacturing: Many 3D printed objects are fragile compared to traditional manufacturing processes, except those made of metal with effective mechanical properties. This is mainly because the components are created layer by layer, which reduces their strength.

Expensive at scale: Generally, carrying out effective manufacturing processes through 3D printing is more expensive due to the absence of economies of scale.

Accuracy limitations: The precision of a printed object is determined by the equipment and process used. Certain printers have low tolerances and may deviate from the CAD. Post-processing may help, but it is essential to note that 3D printed products are not always accurate.

Postprocessing requirements: The majority of 3D printed components need post-processing. This includes smoothening and heat treatments to achieve quality standards [164].

Consumes a great deal of energy to operate: 3D printing consumes 100 times the energy conventional manufacturing consumes. Unquestionably, 3D printing contributes to a greener world by enabling the reuse of materials such as plastic. However, it operates on nonrenewable energy. As a result, the only method to overcome this constraint is to use 3D printing for small-batch manufacturing.

The 3D printer is expensive: Even a single 3D printer may cost hundreds to thousands of dollars. Not only that but the software and material needs would be additional. That is why, even 40 years after its inception, many firms have yet to realize the benefits of 3D printing.

3D production requires skilled users for operation: Specific abilities must be acquired before printing the first 3D printed product. Before starting to 3D print, the operator must understand how to deal with CAD files and get familiar with some of the hardware settings on 3D printers.

Consider it slow for mass production: With the introduction of new 3D printers, this restriction of the 3D printer is being addressed. However, mass production would need more time than conventional manufacturing. Additionally, the time required will vary according to the intricacy of the design, the size of the product, and a variety of other factors. Certain 3D printing technologies may need extensive post-processing. If not done correctly, it may potentially ruin the 3D printed object.

3D printing is a possibility for piracy: Piracy is a significant setback for enterprises. Copying a design and selling it at a lesser price might result in the demise of the original firm. 3D printing simplifies the process even more. All that is required to duplicate the design is the file's blueprint. Once the 3D design is complete, printing as many 3D pieces as possible is feasible. Additionally, patent infringement lawsuits may become a significant issue.

Toxic gases can cause health problems: 3D printers also expose humans to harmful pollutants. 3D printers used behind closed doors pose an even more significant threat. Researchers revealed that hazardous particles might be released into the air during printing. These particles may get lodged in users' bloodstreams and induce congestion in their respiratory system, resulting in cancer and other severe disorders.

Limited printing materials: Unable to choose the required material for printing immediately highlights the limitations of 3D printing. While the number of material options rises in lockstep with demand, they remain far fewer than those accessible in conventional production. Plastic is the most often used 3D printing material due to its inexpensive cost and ease of manipulation, owing to its lower melting point. Limited material options make it difficult to expand the number of applications for 3D printing.

3D printing's texture and color limitations: The textures and color possibilities accessible via conventional manufacturing are still more spectacular than what we accomplish with 3D printing. Users must make tradeoffs in terms of color and completed textures when 3D printing; otherwise, postprocessing procedures must be used, which may be time-consuming and not compatible with all 3D printing processes. One may be unhappy with the cheaper 3D printers' surface quality outcomes. Even the most expensive ones may not produce the same quality finish as other techniques. Again, printing on wood and ceramics presents significant challenges.

The 3D printer's limited construction area: One is limited to printing items no prominent than the build area of the 3D printer. These 3D printing restrictions are acutely felt in small-scale 3D printing service centers. These centers must establish the order sizes they accept and refuse orders that exceed the stipulated sizes. Larger 3D printers are currently in use, but they are not affordable to everyone. Even if one were able to get it, covering the extra expenditures would be difficult on a shoestring budget. While a modest room-sized home has been 3D printed, larger multistory structures remain difficult.

3D printing contributes to job scarcity: Job loss is a serious issue, and every country's government is worried about it just as much as its residents. The need for labor is reducing dramatically as a result of automated technologies. The manufacturing sector is a significant employer of workers on a big scale. Without a doubt, the technology significantly reduces human mistakes. However, denying laborers work will not be justified, resulting in economic catastrophes [113].

1.6 Future and conclusions

The primary benefits of 3D printing are design flexibility, mass customization, and constructing complex structures with minimum waste. Due to its low cost, ease of usage, and rapid execution, fused deposition modeling is extensively used. Compared to other methods, it is primarily for quick prototyping, with the printed objects having worse tensile properties and quality. Inkjet printing is a quick approach for 3D manufacturing ceramic suspensions, although it requires further heat processing. Larger structures, such as homes, are printed utilizing contour crafting based on material extrusion. Stereolithography is a groundbreaking 3D printing technology mainly employed with photopolymers and capable of producing very complex objects. Nonetheless, it is a lengthy and complex process that is confined by available resources.

Thermosetting polymers are the most often utilized fast prototyping material. The addition of fibers and nanoparticles to polymers enhances the composite's tensile properties, enabling it to be used as a functional material. Ceramics can design high strength and complicated structures for various applications, including tissue engineering. Therefore, the primary impediment is the scarcity of products suitable for 3D printing ceramics with more precise microstructure and composition control. Additionally, the uptake of 3D printing by the construction industry has been slow. Recent improvements have focused on developing a concrete mixture with superior flowability, processing convenience, mechanical behavior, and aesthetics.

3D printing has created new opportunities for developing smart materials in a range of disciplines, and its potential seems to be endless. It has contributed to modern biomaterials research and innovation, particularly in prototyping complex and customized products with patient-specific needs. It does, however, face challenges such as a supply shortage and regulatory problems. The aviation industry has invested in 3D printing to create customized components with increased strength-to-weight ratios and facilitate on-demand manufacturing and aircraft service. However, limited materials and high prices and the inconsistent quality of 3D printed components impede the adoption of 3D printing in the aviation industry. 3D printing is still in its infancy in the building industry, with just a few big projects worldwide. In comparison to standard techniques, the key drawback is the high cost and decreased mechanical performance. On the other hand, construction automation enabled by 3D printing ensures labor-free building on the moon.

While 3D printing has several benefits, it also has many limitations that would need greater study and development before this technology could be utilized in a wide variety of businesses. The void formation between layers increases porosity, lowering mechanical performance. Another key issue with 3D printing is its anisotropic behavior, which results in mechanical behavior that is different in the vertical direction than in the horizontal direction. Because CAD is based on the tessellation concept, which estimates the design, converting CAD to a 3D printed object usually results in defects and inconsistencies, most notably in curvatures.

While few difficulties have been addressed via intensive research and materials and technologies, there is still much room for progress. While 3D printing is a game-changer for new products and specialized applications, it needs more study to compete with traditional manufacturing processes in the large-scale manufacture of essential commodities. Nevertheless, 3D printing has made enormous strides in the intervening years. Global investment and research will rapidly transition current production techniques to 3D printing in the not-too-distant future.

References

[1] Additive manufacturing – General principles. Available: https://standards.iteh.ai/catalog/standards/iso/a31c7a1c-bb3c-4b56-acb9-33e9602e9a2b/iso-prf-17296-1. [Accessed: 07-Dec-2021].

[2] P. Holzmann, R.J. Breitenecker, A.A. Soomro, E.J. Schwarz, User entrepreneur business models in 3D printing, J. Manuf. Technol. Manag. 28 (1) (2017) 75–94, https://doi.org/10.1108/JMTM-12-2015-0115/FULL/PDF.

[3] A. Hayes, 3D Printing. Available: https://www.investopedia.com/terms/1/3d-printing.asp. [Accessed: 07-Dec-2021].

[4] B. Berman, 3-D printing: the new industrial revolution, Bus. Horiz. 55 (2) (March, 2012) 155–162, https://doi.org/10.1016/J.BUSHOR.2011.11.003.

[5] A. Pîrjan, D.-M. Petroşanu, The impact of 3D printing technology on the society and economy, J. Inf. Syst. Oper. Manag. 7 (2) (2013) 360–370.

[6] H. Kodama, Automatic method for fabricating a three-dimensional plastic model with photo-hardening polymer, Rev. Sci. Instrum. 52 (11) (June, 1998) 1770–1773, https://doi.org/10.1063/1.1136492.

[7] The Evolution Of 3D Printing. Available: https://blog.thomasnet.com/evolution-of-3d-printing. [Accessed: 07-Dec-2021].

[8] Additive Manufacturing 101. Available: https://www.thomasnet.com/insights/an-overview-of-additive-manufacturing/?_ga=2.143833776.449976833.1623753945-1273053483.1623753945. [Accessed: 07-Dec-2021].

[9] The Disruptive Evolution Of 3D Printing. Available: https://www.manufacturing.net/industry40/blog/13104671/the-disruptive-evolution-of-3d-printing. [Accessed: 07-Dec-2021].

[10] $4.1 Billion Industry Forecast In Crazy 3D Printing Stock Market. Available: https://www.forbes.com/sites/tjmccue/2015/07/30/4-1-billion-industry-forecast-in-crazy-3d-printing-stock-market/?sh=3755f4121631. [Accessed: 07-Dec-2021].

[11] How Manufacturers, Engineers Are Stepping Up to Solve COVID-19 Challenges. Available: https://www.thomasnet.com/insights/how-manufacturers-engineers-are-stepping-up-to-solve-covid-19-challenges/?_ga=2.177902880.449976833.1623753945-1273053483.1623753945. [Accessed: 07-Dec-2021].

[12] The Role of Computer-Aided Design (CAD) in 3D Printing. Available: https://monroeengineering.com/blog/the-role-of-computer-aided-design-cad-in-3d-printing/. [Accessed: 07-Dec-2021].

[13] S. Tornincasa, F. Di Monaco, The future and the evolution of CAD, in: Proceedings of the 14th International Research/expert Conference: Trends in the Development of Machinery and Associated Technology, vol. 1, 2010, pp. 11–18, 1.

[14] T. Chen, P. Egan, F. Stöckli, K. Shea, Studying the impact of incorporating an additive manufacturing based design Exercise in a large, first year technical drawing and CAD Course, Proc. ASME Des. Eng. Tech. Conf. 3 (January, 2016), https://doi.org/10.1115/DETC2015-47312.

[15] S. Junk, R. Matt, New approach to introduction of 3D digital technologies in design education, Proc. CIRP 36 (January, 2015) 35–40, https://doi.org/10.1016/J.PROCIR.2015.01.045.

[16] S.A.M. Tofail, E.P. Koumoulos, A. Bandyopadhyay, S. Bose, L. O'Donoghue, C. Charitidis, Additive manufacturing: scientific and technological challenges, market uptake and opportunities, Mater. Today 21 (1) (January, 2018) 22–37, https://doi.org/10.1016/J.MATTOD.2017.07.001.

[17] J.W. Stansbury, M.J. Idacavage, 3D printing with polymers: challenges among expanding options and opportunities, Dent. Mater. 32 (1) (January, 2016) 54–64, https://doi.org/10.1016/J.DENTAL.2015.09.018.

[18] Z.X. Low, Y.T. Chua, B.M. Ray, D. Mattia, I.S. Metcalfe, D.A. Patterson, Perspective on 3D printing of separation membranes and comparison to related unconventional fabrication techniques, J. Membr. Sci. 523 (February, 2017) 596–613, https://doi.org/10.1016/J.MEMSCI.2016.10.006.

[19] C. Lee Ventola, Medical applications for 3D printing: current and projected uses, Pharm. Ther. 39 (10) (October, 2014) 704.

[20] C. Silbernagel, "Additive Manufacturing : What is material jetting?" Available: http://canadamakes.ca/what-is-material-jetting/. [Accessed: 07-Dec-2021].

[21] The Types of 3D Printing Technology of 2021. Available: https://all3dp.com/1/types-of-3d-printers-3d-printing-technology/. [Accessed: 07-Dec-2021].

[22] I. Gibson, D. Rosen, B. Stucker, M. Khorasani, Binder jetting, Addit. Manuf. Technol. (2021) 237–252, https://doi.org/10.1007/978-3-030-56127-7_8.

[23] I. Chiulan, A.N. Frone, C. Brandabur, D.M. Panaitescu, Recent advances in 3D printing of aliphatic Polyesters, Bioengineering 5 (1) (December, 2017), https://doi.org/10.3390/BIOENGINEERING5010002, 2018, Vol. 5, Page 2.

[24] S.C. Park, H.S. Bang, W.J. Seong, "Effects of material properties on angular distortion in Wire Arc additive manufacturing: experimental and computational Analyses," Mater 13 (6) (March, 2020) 1399, https://doi.org/10.3390/MA13061399, 2020, Vol. 13, Page 1399.

[25] I. Gibson, D. Rosen, B. Stucker, Directed energy deposition processes, Addit. Manuf. Technol. (2015) 245–268, https://doi.org/10.1007/978-1-4939-2113-3_10.

[26] "What is Directed Energy Deposition (DED)?" Available: https://www.twi-global.com/technical-knowledge/faqs/directed-energy-deposition. [Accessed: 07-Dec-2021].

[27] W.Y. Yeong, C.K. Chua, Bioprinting: Principles and Applications, vol. 1, World Scientific Publishing Co Inc, 2014.

[28] L. Moroni, et al., Biofabrication strategies for 3D in vitro models and regenerative medicine, Nat. Rev. Mater. 3 (5) (April, 2018) 21–37, https://doi.org/10.1038/s41578-018-0006-y, 2018 35.

[29] J. Gopinathan, I. Noh, Recent trends in bioinks for 3D printing, Biomater. Res. 22 (1) (April, 2018) 1–15, https://doi.org/10.1186/S40824-018-0122-1, 2018 221.

[30] D. Thomas, Z.M. Jessop, I.S. Whitaker, 3D Bioprinting for Reconstructive Surgery: Techniques and Applications, Woodhead Publishing, 2017.

[31] H.-W. Kang, S.J. Lee, I.K. Ko, C. Kengla, J.J. Yoo, A. Atala, A 3D bioprinting system to produce human-scale tissue constructs with structural integrity, Nat. Biotechnol. 34 (3) (2016) 312–319.

[32] S.C. Lee, G. Gillispie, P. Prim, S.J. Lee, Physical and chemical factors influencing the printability of hydrogel-based extrusion bioinks, Chem. Rev. 120 (19) (October, 2020) 10834–10886, https://doi.org/10.1021/ACS.CHEMREV.0C00015.

[33] A. Panwar, L.P. Tan, Current Status of bioinks for micro-extrusion-based 3D bioprinting, Molecules 21 (6) (May, 2016) 685, https://doi.org/10.3390/MOLECULES21060685.

[34] J.H. Kim, J.J. Yoo, S.J. Lee, Three-dimensional cell-based bioprinting for soft tissue regeneration, Tissue Eng. Regen. Med. 13 (6) (December, 2016) 647–662, https://doi.org/10.1007/S13770-016-0133-8, 2016 136.
[35] S. V Murphy, A. Atala, 3D bioprinting of tissues and organs, Nat. Biotechnol. 32 (8) (2014) 773–785, https://doi.org/10.1038/nbt.2958.
[36] J. An, J.E.M. Teoh, R. Suntornnond, C.K. Chua, Design and 3D printing of scaffolds and tissues, Engineering 1 (2) (June, 2015) 261–268, https://doi.org/10.15302/J-ENG-2015061.
[37] D. Seliktar, D. Dikovsky, E. Napadensky, Bioprinting and tissue engineering: recent advances and future perspectives, Isr. J. Chem. 53 (9–10) (October, 2013) 795–804, https://doi.org/10.1002/IJCH.201300084.
[38] J.P. Kruth, G. Levy, F. Klocke, T.H.C. Childs, Consolidation phenomena in laser and powder-bed based layered manufacturing, CIRP Ann 56 (2) (January, 2007) 730–759, https://doi.org/10.1016/J.CIRP.2007.10.004.
[39] S.S. Alghamdi, S. John, N.R. Choudhury, N.K. Dutta, Additive manufacturing of polymer materials: progress, promise and challenges, Polymer 13 (5) (February, 2021) 753, https://doi.org/10.3390/POLYM13050753, 2021, Vol. 13, Page 753.
[40] D.J. Horst, C.A. Duvoisin, R. de Almeida Vieira, Additive manufacturing at Industry 4.0: a review, Int. J. Eng. Tech. Res. 8 (8) (2018).
[41] J.H. Martin, B.D. Yahata, J.M. Hundley, J.A. Mayer, T.A. Schaedler, T.M. Pollock, 3D printing of high-strength aluminium alloys, Nature 549 (7672) (September, 2017) 365–369, https://doi.org/10.1038/nature23894.
[42] L. Hitzler, et al., Additive manufacturing of cobalt-based dental alloys: analysis of microstructure and physicomechanical properties, Adv. Mater. Sci. Eng. 2018 (2018), https://doi.org/10.1155/2018/8213023.
[43] L.E. Murr, Frontiers of 3D printing/additive manufacturing: from human organs to aircraft fabrication, J. Mater. Sci. Technol. 32 (10) (October, 2016) 987–995, https://doi.org/10.1016/J.JMST.2016.08.011.
[44] T. DebRoy, et al., Additive manufacturing of metallic components – process, structure and properties, Prog. Mater. Sci. 92 (March, 2018) 112–224, https://doi.org/10.1016/J.PMATSCI.2017.10.001.
[45] E. Uhlmann, R. Kersting, T.B. Klein, M.F. Cruz, A.V. Borille, Additive manufacturing of titanium alloy for aircraft components, Procedia CIRP 35 (January, 2015) 55–60, https://doi.org/10.1016/J.PROCIR.2015.08.061.
[46] F. Trevisan, et al., Additive manufacturing of titanium alloys in the biomedical field: processes, properties and applications, J. Appl. Biomater. Funct. Mater. 16 (2) (April, 2018) 57–67, https://doi.org/10.5301/jabfm.5000371.
[47] M.A. Caminero, J.M. Chacón, I. García-Moreno, G.P. Rodríguez, Impact damage resistance of 3D printed continuous fibre reinforced thermoplastic composites using fused deposition modelling, Compos. B Eng. 148 (September, 2018) 93–103, https://doi.org/10.1016/J.COMPOSITESB.2018.04.054.
[48] J.R.C. Dizon, A.H. Espera, Q. Chen, R.C. Advincula, Mechanical characterization of 3D-printed polymers, Addit. Manuf. 20 (March, 2018) 44–67, https://doi.org/10.1016/J.ADDMA.2017.12.002.
[49] X. Wang, M. Jiang, Z. Zhou, J. Gou, D. Hui, 3D printing of polymer matrix composites: a review and prospective, Compos. B Eng. 110 (February, 2017) 442–458, https://doi.org/10.1016/J.COMPOSITESB.2016.11.034.
[50] F. Baldassarre, F. Ricciardi, The additive manufacturing in the Industry 4.0 Era: the case of an Italian FabLab, J. Emerg. Trends Mark. Manag. 1 (1) (2017) 105–115.
[51] R. Gmeiner, et al., Additive manufacturing of bioactive glasses and silicate bioceramics, J. Ceram. Sci. Technol. 6 (2) (2015) 75–86.

[52] D. Owen, et al., 3D printing of ceramic components using a customized 3D ceramic printer, Prog. Addit. Manuf. 3 (1–2) (June, 2018) 3–9, https://doi.org/10.1007/S40964-018-0037-3/FIGURES/6.
[53] A. Zocca, P. Lima, J. Günster, LSD-based 3D printing of alumina ceramics, J. Ceram. Sci. Technol. 8 (1) (2017) 141–148.
[54] T. Ianko, S. Panov, O. Sushchyns'ky, M. Pylypenko, O. Dmytrenko, Zirconium Alloy Powders for Manufacture of 3d Printed Articles Used in Nuclear Power Industry, Вопросы атомной науки и техники, 2018.
[55] W. Hao, Y. Liu, H. Zhou, H. Chen, D. Fang, Preparation and characterization of 3D printed continuous carbon fiber reinforced thermosetting composites, Polym. Test. 65 (February, 2018) 29–34, https://doi.org/10.1016/J.POLYMERTESTING.2017.11.004.
[56] T.P. Sathishkumar, S. Satheeshkumar, J. Naveen, Glass fiber-reinforced polymer composites – a review, J. Reinforc. Plast. Compos. 33 (13) (April, 2014) 1258–1275, https://doi.org/10.1177/0731684414530790.
[57] Z. Liu, et al., Modification of glass fiber surface and glass fiber reinforced polymer composites challenges and opportunities: from organic chemistry perspective, Curr. Org. Chem. 19 (11) (2015) 991–1010.
[58] J.Y. Lee, J. An, C.K. Chua, Fundamentals and applications of 3D printing for novel materials, Appl. Mater. Today 7 (June, 2017) 120–133, https://doi.org/10.1016/J.APMT.2017.02.004.
[59] H.J. Van, Additive manufacturing of shape memory alloy, Shape Mem. superelasticity 4 (2) (2018).
[60] Y. Yang, Y. Chen, Y. Wei, Y. Li, 3D printing of shape memory polymer for functional part fabrication, Int. J. Adv. Manuf. Technol. 84 (9) (September, 2015) 2079–2095, https://doi.org/10.1007/S00170-015-7843-2, 2015 849.
[61] A. Perrot, D. Rangeard, E. Courteille, 3D printing of earth-based materials: processing aspects, Construct. Build. Mater. 172 (May, 2018) 670–676, https://doi.org/10.1016/J.CONBUILDMAT.2018.04.017.
[62] Everything About a 3D Printing Pen. Available: https://manufactur3dmag.com/everything-about-a-3d-printing-pen/. [Accessed: 09-Dec-2021].
[63] A. Goyanes, et al., Automated therapy preparation of isoleucine formulations using 3D printing for the treatment of MSUD: first single-centre, prospective, crossover study in patients, Int. J. Pharm. 567 (August, 2019) 118497, https://doi.org/10.1016/J.IJPHARM.2019.118497.
[64] I. Seoane-Viaño, et al., 3D printed tacrolimus suppositories for the treatment of ulcerative colitis, Asian J. Pharm. Sci. 16 (1) (January, 2021) 110–119, https://doi.org/10.1016/J.AJPS.2020.06.003.
[65] J. Elbl, J. Gajdziok, J. Kolarczyk, 3D printing of multilayered orodispersible films with in-process drying, Int. J. Pharm. 575 (February, 2020) 118883, https://doi.org/10.1016/J.IJPHARM.2019.118883.
[66] H.G. Yi, et al., A 3D-printed local drug delivery patch for pancreatic cancer growth suppression, J. Contr. Release 238 (September, 2016) 231–241, https://doi.org/10.1016/J.JCONREL.2016.06.015.
[67] S.A. Khaled, J.C. Burley, M.R. Alexander, J. Yang, C.J. Roberts, 3D printing of tablets containing multiple drugs with defined release profiles, Int. J. Pharm. 494 (2) (October, 2015) 643–650, https://doi.org/10.1016/J.IJPHARM.2015.07.067.
[68] H.W. Cho, S.H. Baek, B.J. Lee, H.E. Jin, Orodispersible polymer films with the poorly water-soluble drug, olanzapine: hot-melt pneumatic extrusion for single-process 3D printing, Pharm. Times 12 (8) (July, 2020) 692, https://doi.org/10.3390/PHARMACEUTICS12080692, 2020, Vol. 12, Page 692.

[69] E. Sjöholm, et al., 3D-Printed veterinary dosage forms—a comparative study of three semi-solid extrusion 3D printers, Pharm. Times 12 (12) (December, 2020) 1239, https://doi.org/10.3390/PHARMACEUTICS12121239, 2020, Vol. 12, Page 1239.

[70] E. Naseri, C. Cartmell, M. Saab, R.G. Kerr, A. Ahmadi, Development of 3D printed drug-eluting scaffolds for preventing piercing infection, Pharm. Times 12 (9) (September, 2020) 901, https://doi.org/10.3390/PHARMACEUTICS12090901, 2020, Vol. 12, Page 901.

[71] E.G. Andriotis, G.K. Eleftheriadis, C. Karavasili, D.G. Fatouros, Development of bio-active patches based on Pectin for the treatment of Ulcers and wounds using 3D-bioprinting technology, Pharm. Times 12 (1) (January, 2020) 56, https://doi.org/10.3390/PHARMACEUTICS12010056, 2020, Vol. 12, Page 56.

[72] J. Holländer, R. Hakala, J. Suominen, N. Moritz, J. Yliruusi, N. Sandler, 3D printed UV light cured polydimethylsiloxane devices for drug delivery, Int. J. Pharm. 544 (2) (June, 2018) 433–442, https://doi.org/10.1016/J.IJPHARM.2017.11.016.

[73] N. Noor, A. Shapira, R. Edri, I. Gal, L. Wertheim, T. Dvir, 3D printing of personalized thick and perfusable cardiac patches and hearts, Adv. Sci. 6 (11) (2019) 1900344.

[74] A. Isaacson, S. Swioklo, C.J. Connon, 3D bioprinting of a corneal stroma equivalent, Exp. Eye Res. 173 (August, 2018) 188–193, https://doi.org/10.1016/J.EXER.2018.05.010.

[75] L.J. Pourchet, et al., Human skin 3D bioprinting using scaffold-free approach, Adv. Healthc. Mater. 6 (4) (February, 2017) 1601101, https://doi.org/10.1002/ADHM.201601101.

[76] M.S. Mannoor, et al., 3D printed bionic ears, Nano Lett. 13 (6) (June, 2013) 2634–2639, https://doi.org/10.1021/NL4007744/SUPPL_FILE/NL4007744_SI_003.AVI.

[77] S.N. Kurenov, C. Ionita, D. Sammons, T.L. Demmy, Three-dimensional printing to facilitate anatomic study, device development, simulation, and planning in thoracic surgery, J. Thorac. Cardiovasc. Surg. 149 (4) (April, 2015) 973–979, https://doi.org/10.1016/J.JTCVS.2014.12.059, e1.

[78] N. Van Assche, et al., Accuracy of implant placement based on pre-surgical planning of three-dimensional cone-beam images: a pilot study, J. Clin. Periodontol. 34 (9) (September, 2007) 816–821, https://doi.org/10.1111/J.1600-051X.2007.01110.X.

[79] A.M. Sanna, L. Molly, D. van Steenberghe, Immediately loaded CAD-CAM manufactured fixed complete dentures using flapless implant placement procedures: a cohort study of consecutive patients, J. Prosthet. Dent 97 (6) (June, 2007) 331–339, https://doi.org/10.1016/S0022-3913(07)60021-3.

[80] P.B. Tardieu, L. Vrielinck, E. Escolano, M. Henne, A.-L. Tardieu, Computer-assisted implant placement: scan template, simplant, surgiguide, and SAFE system, Int. J. Periodontics Restor. Dent. 27 (2) (2007).

[81] D. Detolla, S. Andreana, A. Patra, R. Buhite, B. Comella, The role of the finite element model in dental implants, J. Oral Implantol. 26 (2000) 77–81, https://doi.org/10.1563/1548-1336(2000)026.

[82] T.V. Flügge, K. Nelson, R. Schmelzeisen, M.C. Metzger, Three-dimensional plotting and printing of an implant drilling guide: simplifying guided implant surgery, J. Oral Maxillofac. Surg. 71 (8) (August, 2013) 1340–1346, https://doi.org/10.1016/J.JOMS.2013.04.010.

[83] J. Chen, Z. Zhang, X. Chen, C. Zhang, G. Zhang, Z. Xu, Design and manufacture of customized dental implants by using reverse engineering and selective laser melting technology, J. Prosthet. Dent. 112 (5) (November, 2014) 1088–1095.e1, https://doi.org/10.1016/J.PROSDENT.2014.04.026.

[84] Invisalign - How does it work?. Available: https://www.youtube.com/watch?v=60bJYGDaqx0. [Accessed: 09-Dec-2021].

[85] A. Dawood, S. Tanner, I. Hutchison, Computer guided surgery for implant placement and dental rehabilitation in a patient undergoing sub-total mandibulectomy and microvascular free Flap reconstruction, J. Oral Implantol. 39 (4) (August, 2013) 497–502, https://doi.org/10.1563/AAID-JOI-D-11-00142.

[86] S. Akyalcin, B.E. Cozad, J.D. English, C.D. Colville, S. Laman, Diagnostic accuracy of impression-free digital models, Am. J. Orthod. Dentofacial Orthop. 144 (6) (December, 2013) 916–922, https://doi.org/10.1016/J.AJODO.2013.04.024.

[87] W.S. Lin, J.C. Chou, M.J. Metz, B.T. Harris, D. Morton, Use of intraoral digital scanning for a CAD/CAM-fabricated milled bar and superstructure framework for an implant-supported, removable complete dental prosthesis, J. Prosthet. Dent 113 (6) (June, 2015) 509–515, https://doi.org/10.1016/J.PROSDENT.2015.01.014.

[88] A. Ender, A. Mehl, Accuracy of complete-arch dental impressions: a new method of measuring trueness and precision, J. Prosthet. Dent 109 (2) (February, 2013) 121–128, https://doi.org/10.1016/S0022-3913(13)60028-1.

[89] F. Ciuffolo, et al., Rapid prototyping: a new method of preparing trays for indirect bonding, Am. J. Orthod. Dentofacial Orthop. 129 (1) (January, 2006) 75–77, https://doi.org/10.1016/J.AJODO.2005.10.005.

[90] Y. Xiong, C. Qian, J. Sun, Fabrication of porous titanium implants by three-dimensional printing and sintering at different temperatures, Dent. Mater. J. 31 (5) (October, 2012) 815–820, https://doi.org/10.4012/DMJ.2012-065.

[91] D.C. Hatcher, C. Dial, C. Mayorga, Cone beam CT for pre-surgical assessment of implant sites, CDA 31 (11) (2003) 825–834.

[92] B. O. G., G. V. S., S. B. K., D. R. C., G. S. A., G. P. S., 3D printing and pharmaceutical manufacturing: opportunities and challenges, Int. J. Bioassays 5 (1) (January, 2016) 4723–4738, https://doi.org/10.21746/IJBIO.2016.01.006.

[93] N. Shahrubudin, T.C. Lee, R. Ramlan, An overview on 3D printing technology: technological, materials, and applications, Procedia Manuf. 35 (January, 2019) 1286–1296, https://doi.org/10.1016/J.PROMFG.2019.06.089.

[94] S. Kotta, A. Nair, N. Alsabeelah, 3D printing technology in drug delivery: recent progress and application, Curr. Pharmaceut. Des. 24 (42) (2018) 5039–5048, https://doi.org/10.2174/1381612825666181206123828.

[95] 3D Printing Houses. Available: https://www.asme.org/topics-resources/content/3d-printing-houses. [Accessed: 10-Dec-2021].

[96] M.D. Symes, et al., Integrated 3D-printed reactionware for chemical synthesis and analysis, Nat. Chem. 4 (5) (April, 2012) 349–354, https://doi.org/10.1038/nchem.1313.

[97] J.L. Erkal, et al., 3D printed microfluidic devices with integrated versatile and reusable electrodes, Lab Chip 14 (12) (May, 2014) 2023–2032, https://doi.org/10.1039/C4LC00171K.

[98] P.J. Kitson, M.H. Rosnes, V. Sans, V. Dragone, L. Cronin, Configurable 3D-Printed millifluidic and microfluidic 'lab on a chip' reactionware devices, Lab Chip 12 (18) (September, 2012) 3267–3271, https://doi.org/10.1039/C2LC40761B/.

[99] V. Dragone, V. Sans, M.H. Rosnes, P.J. Kitson, L. Cronin, 3D-printed devices for continuous-flow organic chemistry, Beilstein J. Org. Chem. 9 (1) (May, 2013) 951–959, https://doi.org/10.3762/BJOC.9.109.

[100] D.L. Elbert, Bottom-up tissue engineering, Curr. Opin. Biotechnol. 22 (5) (October, 2011) 674–680, https://doi.org/10.1016/J.COPBIO.2011.04.001.

[101] K. Jakab, C. Norotte, F. Marga, K. Murphy, G. Vunjak-Novakovic, G. Forgacs, Tissue engineering by self-assembly and bio-printing of living cells, Biofabrication 2 (2) (June, 2010) 022001, https://doi.org/10.1088/1758-5082/2/2/022001.

[102] K. Jakab, et al., Tissue engineering by self-assembly of cells printed into topologically defined structures, Tissue Eng. 14 (3) (March, 2008) 413–421, https://doi.org/10.1089/TEA.2007.0173.

[103] X. Wang, Y. Yan, R. Zhang, Rapid prototyping as a tool for manufacturing bioartificial livers, Trend. Biotechnol. 25 (11) (November, 2007) 505–513, https://doi.org/10.1016/J.TIBTECH.2007.08.010.

[104] B. Guillotin, F. Guillemot, Cell patterning technologies for organotypic tissue fabrication, Trend. Biotechnol. 29 (4) (April, 2011) 183–190, https://doi.org/10.1016/J.TIBTECH.2010.12.008.

[105] S.S. Kim, et al., Survival and function of hepatocytes on a novel three-dimensional synthetic biodegradable polymer scaffold with an intrinsic network of channels, Ann. Surg. 228 (1) (July, 1998) 8, https://doi.org/10.1097/00000658-199807000-00002.

[106] L.G. Griffith, B. Wu, M.J. Cima, M.J. Powers, B. Chaignaud, J.P. Vacanti, In vitro Organogenesis of liver Tissuea, Ann. N. Y. Acad. Sci. 831 (1) (December, 1997) 382–397, https://doi.org/10.1111/J.1749-6632.1997.TB52212.X.

[107] J.S. Miller, et al., Rapid casting of patterned vascular networks for perfusable engineered three-dimensional tissues, Nat. Mater. 11 (9) (July, 2012) 768–774, https://doi.org/10.1038/nmat3357.

[108] C. Norotte, F.S. Marga, L.E. Niklason, G. Forgacs, Scaffold-free vascular tissue engineering using bioprinting, Biomaterials 30 (30) (October, 2009) 5910–5917, https://doi.org/10.1016/J.BIOMATERIALS.2009.06.034.

[109] M. Porteus, L. Stem, Translating the Lessons from Gene therapy to the development of regenerative medicine, Mol. Ther. 19 (3) (March, 2011) 439–441, https://doi.org/10.1038/MT.2011.14.

[110] C. Mavroidis, et al., Patient specific ankle-foot orthoses using rapid prototyping, J. NeuroEng. Rehabil. 8 (1) (January, 2011) 1–11, https://doi.org/10.1186/1743-0003-8-1/FIGURES/10.

[111] P. Boileau, D.J. Watkinson, A.M. Hatzidakis, F. Balg, Grammont reverse prosthesis: design, rationale, and biomechanics, J. Shoulder Elbow Surg. 14 (1) (January, 2005) S147–S161, https://doi.org/10.1016/J.JSE.2004.10.006.

[112] H. Li, X. Qu, Y. Mao, K. Dai, Z. Zhu, Custom acetabular cages offer stable fixation and improved hip scores for revision THA with severe bone defects, Clin. Orthop. Relat. Res. 474 (3) (2016) 731–740.

[113] M.A. Alhnan, T.C. Okwuosa, M. Sadia, K.W. Wan, W. Ahmed, B. Arafat, Emergence of 3D printed dosage forms: opportunities and challenges, Pharm. Res. 33 (8) (August, 2016) 1817–1832, https://doi.org/10.1007/S11095-016-1933-1/FIGURES/1.

[114] M. McGurk, A.A. Amis, P. Potamianos, N.M. Goodger, Rapid prototyping techniques for anatomical modelling in medicine, Ann. R. Coll. Surg. Engl. 79 (3) (1997) 169.

[115] C.E. Dombroski, M.E.R. Balsdon, A. Froats, The use of a low cost 3D scanning and printing tool in the manufacture of custom-made foot orthoses: a preliminary study, BMC Res. Notes 7 (1) (July, 2014) 1–4, https://doi.org/10.1186/1756-0500-7-443/FIGURES/2.

[116] H.H. Malik, et al., Three-dimensional printing in surgery: a review of current surgical applications, J. Surg. Res. 199 (2) (December, 2015) 512–522, https://doi.org/10.1016/J.JSS.2015.06.051.

[117] F.A. Khan, et al., 3D printing technology in customized drug delivery system: current state of the Art, prospective and the challenges, Curr. Pharmaceut. Des. 24 (42) (January, 2019) 5049–5061, https://doi.org/10.2174/1381612825666190110153742.

[118] A. Dawood, B.M. Marti, V. Sauret-Jackson, A. Darwood, 3D printing in dentistry, BDJ 219 (11) (2015) 521–529, 2015.

[119] K. Surmen, F. Ortes, Y.Z. Arslan, Design and production of subject specific insole using reverse engineering and 3D printing technology, Int. J. Eng. Sci. Inven. 5 (12) (2016) 11–15.

[120] C. Damianou, M. Giannakou, C. Yiallouras, G. Menikou, The role of three-dimensional printing in magnetic resonance imaging-guided focused ultrasound surgery, Digit. Med. 4 (1) (2018) 22, https://doi.org/10.4103/DIGM.DIGM_48_17.

[121] P. Canzi, et al., From CT scanning to 3D printing technology: a new method for the preoperative planning of a transcutaneous bone-conduction hearing device, Acta Otorhinolaryngol. Ital. 38 (3) (June, 2018) 251, https://doi.org/10.14639/0392-100X-1625.

[122] B. Zhang, et al., 3D bioprinting: an emerging technology full of opportunities and challenges, Bio-Design Manuf. 1 (1) (February, 2018) 2–13, https://doi.org/10.1007/S42242-018-0004-3.

[123] R. Levato, J. Visser, J.A. Planell, E. Engel, J. Malda, M.A. Mateos-Timoneda, Biofabrication of tissue constructs by 3D bioprinting of cell-laden microcarriers, Biofabrication 6 (3) (July, 2014) 035020, https://doi.org/10.1088/1758-5082/6/3/035020.

[124] J.A. Schrader, H.A. Kratsch, W.R. Graves, Bioplastic Container Cropping Systems: Green Technology for the Green Industry, Sustain. Hort. Res. Consortium, Ames, IA, 2016.

[125] R. Mathur, 3D printing in architecture, Int. J. Innov. Sci. Eng. Technol. 3 (7) (2016) 583–591.

[126] H. Yin, M. Qu, H. Zhang, Y.C. Lim, 3D printing and buildings: a technology review and future Outlook, Technol. Archit. Des. 2 (1) (2018) 94–111, https://doi.org/10.1080/24751448.2018.1420968.

[127] J. Chard, M. van Iersel, B. Bugbee, Mini-lysimeters to monitor transpiration and control Drought stress: system design and unique applications, Tech. Instrument. (January, 2010) 1–30.

[128] L. Hess, H. De Kroon, Effects of rooting volume and nutrient availability as an alternative explanation for root self/non-self discrimination, J. Ecol. 95 (2) (March, 2007) 241–251, https://doi.org/10.1111/J.1365-2745.2006.01204.X.

[129] H. Poorter, J. Bühler, D. van Dusschoten, J. Climent, J.A. Postma, Pot size matters: a meta-analysis of the effects of rooting volume on plant growth, Funct. Plant Biol. 39 (11) (2012) 839–850.

[130] I. Biran, A. Eliassaf, The effect of container size and aeration conditions on growth of roots and canopy of woody plants, Sci. Hortic. 12 (4) (June, 1980) 385–394, https://doi.org/10.1016/0304-4238(80)90054-0.

[131] J.D. Ray, T.R. Sinclair, The effect of pot size on growth and transpiration of maize and soybean during water deficit stress, J. Exp. Bot. 49 (325) (1998) 1381–1386.

[132] G. Nikolaou, D. Neocleous, N. Katsoulas, C. Kittas, Irrigation of greenhouse Crops, Horticulturae 5 (1) (January, 2019) 7, https://doi.org/10.3390/HORTICULTURAE5010007.

[133] F. Massetani, G. Savini, D. Neri, Effect of rooting time, pot size and fertigation technique on strawberry plant architecture, J. Berry Res. 4 (4) (2014) 217–224.

[134] A. Coffey, E. Prinsen, M.A.K. Jansen, J. Conway, The UVB photoreceptor UVR8 mediates accumulation of UV-absorbing pigments, but not changes in plant morphology, under outdoor conditions, Plant Cell Environ. 40 (10) (October, 2017) 2250–2260, https://doi.org/10.1111/PCE.13025.

[135] A. Kapanen, E. Schettini, G. Vox, M. Itävaara, Performance and environmental impact of biodegradable films in agriculture: a field study on protected cultivation, J. Polym. Environ. 16 (2) (May, 2008) 109–122, https://doi.org/10.1007/S10924-008-0091-X/TABLES/2.

[136] R. Müller-Xing, Q. Xing, J. Goodrich, Footprints of the sun: memory of UV and light stress in plants, Front. Plant Sci. 5 (SEP) (September, 2014) 474, https://doi.org/10.3389/FPLS.2014.00474/BIBTEX.

[137] E. Elfadly, H. Abd El-Aal, A. Rizk, W. Sobeih, Ambient UV manipulation in greenhouses: plant responses and insect pest management in cucumber, in: VIII International Symposium on Light in Horticulture, vol. 1134, 2016, pp. 343–350.

[138] G.I. Jenkins, The UV-B photoreceptor UVR8: from structure to physiology, Plant Cell 26 (1) (2014) 21–37.

[139] D. Assante, G.M. Cennamo, L. Placidi, 3D printing in education: an european perspective, IEEE Glob. Eng. Educ. Conf. EDUCON 2020 (April, 2020) 1133–1138, https://doi.org/10.1109/EDUCON45650.2020.9125311.
[140] S.S. Horowitz, P.H. Schultz, Printing Space: using 3D printing of digital Terrain models in Geosciences education and research, J. Geosci. Educ. 62 (1) (February, 2018) 138–145, https://doi.org/10.5408/13-031.1.
[141] I. Verner, A. Merksamer, Digital design and 3D printing in technology teacher education, Procedia CIRP 36 (January, 2015) 182–186, https://doi.org/10.1016/J.PROCIR.2015.08.041.
[142] C. Schelly, G. Anzalone, B. Wijnen, J.M. Pearce, Open-source 3-D printing technologies for education: Bringing additive manufacturing to the classroom, J. Vis. Lang. Comput. 28 (June, 2015) 226–237, https://doi.org/10.1016/J.JVLC.2015.01.004.
[143] E. Buehler, et al., Investigating the Implications of 3D printing in special education, ACM Trans. Access. Comput. 8 (3) (March, 2016) 11, https://doi.org/10.1145/2870640.
[144] R.L. Martin, N.S. Bowden, C. Merrill, 3D printing in technology and engineering education, Technol. Eng. Teach. 73 (8) (2014) 30.
[145] A. Stangl, J. Kim, T. Yeh, 3D printed tactile picture books for children with visual Impairments: a design Probe, Proc. 2014 Conf. Interact. Des. Child. (2014), https://doi.org/10.1145/2593968.
[146] F. Dilling, Funktionsweise der 3D-Druck-Technologie, in: Der Einsatz der 3D-Druck-Technologie im Mathematikunterricht, Springer, 2019, pp. 3–4.
[147] I. Witzke, S. Spies, Domain-specific Beliefs of School calculus, J. fur Math. 37 (1) (July, 2016) 131–161, https://doi.org/10.1007/S13138-016-0106-4/FIGURES/6.
[148] D. Tall, How Humans Learn to Think Mathematically: Exploring the Three Worlds of Mathematics, Cambridge University Press, 2013.
[149] T. Asmaria, et al., The 3D printing in material research and medical Physics education and its accuracy study, J. Penelit. Pengemb. Pendidik. Fis. 6 (2) (2020) 227–236.
[150] J. Zhu, J. Yang, C. Tang, Z. Cong, X. Cai, C. Ma, Design and validation of a 3D-printed simulator for endoscopic third ventriculostomy, Child's Nerv. Syst. 36 (4) (April, 2020) 743–748, https://doi.org/10.1007/S00381-019-04421-8/FIGURES/5.
[151] T. Acar, A.B. Karakas, M.A. Ozer, A.M. Koc, F. Govsa, Building three-dimensional intracranial Aneurysm models from 3D-TOF MRA: a validation study, J. Digit. Imag. 32 (6) (December, 2019) 963–970, https://doi.org/10.1007/S10278-019-00256-6/FIGURES/6.
[152] V.P. Bodani, G.E. Breimer, F.A. Haji, T. Looi, J.M. Drake, Development and evaluation of a patient-specific surgical simulator for endoscopic colloid cyst resection, J. Neurosurg. 133 (2) (June, 2019) 521–529, https://doi.org/10.3171/2019.4.JNS183184.
[153] F.W. Grillo, et al., Patient-specific neurosurgical phantom: assessment of visual quality, accuracy, and scaling effects, 3D Print. Med. 4 (1) (March, 2018) 1–9, https://doi.org/10.1186/S41205-018-0025-8, 2018 41.
[154] P. Weinstock, S.P. Prabhu, K. Flynn, D.B. Orbach, E. Smith, Optimizing cerebrovascular surgical and endovascular procedures in children via personalized 3D printing, J. Neurosurg. Pediatr. 16 (5) (November, 2015) 584–589, https://doi.org/10.3171/2015.3.PEDS14677.
[155] Z. Yi, B. He, Y. Liu, S. Huang, W. Hong, Development and evaluation of a craniocerebral model with tactile-realistic feature and intracranial pressure for neurosurgical training, J. Neurointerventional Surg. 12 (1) (January, 2020) 94–97, https://doi.org/10.1136/NEURINTSURG-2019-015008.
[156] A.S. Rose, J.S. Kimbell, C.E. Webster, O.L.A. Harrysson, E.J. Formeister, C.A. Buchman, Multi-material 3D models for temporal bone surgical simulation,

Ann. Otol. Rhinol. Laryngol. 124 (7) (January, 2015) 528–536, https://doi.org/10.1177/0003489415570937.

[157] J.B. Hochman, C. Rhodes, D. Wong, J. Kraut, J. Pisa, B. Unger, Comparison of cadaveric and isomorphic three-dimensional printed models in temporal bone education, Laryngoscope 125 (10) (October, 2015) 2353–2357, https://doi.org/10.1002/LARY.24919.

[158] D. Periard, N. Schaal, M. Schaal, E. Malone, H. Lipson, Printing food, in: 2007 International Solid Freeform Fabrication Symposium, 2007.

[159] E. Malone, H. Lipson, Fab@ Home: the personal desktop fabricator kit, Rapid Prototyp. J. 13 (4) (2007) 245–255.

[160] J. Lipton, et al., Multi-material food printing with complex internal structure suitable for conventional post-processing, in: 2010 International Solid Freeform Fabrication Symposium, 2010.

[161] L. Hao, S. Mellor, O. Seaman, J. Henderson, N. Sewell, M. Sloan, Material characterisation and process development for chocolate additive layer manufacturing, Virtual Phys. Prototyp. 5 (2) (June, 2010) 57–64, https://doi.org/10.1080/17452751003753212.

[162] L. Hao, O. Seaman, S. Mellor, J. Henderson, N. Sewell, M. Sloan, Extrusion behavior of chocolate for additive layer manufacturing, Innov. Dev. Des. Manuf. - Adv. Res. Virtual Rapid Prototyp. (September, 2010) 245–250, doi: 10.1201/9780203859476-45/EXTRUSION-BEHAVIOR-CHOCOLATE-ADDITIVE-LAYER-MANUFACTURING-HAO-SEAMAN-MELLOR-HENDERSON-SEWELL-SLOAN.

[163] Industrial Applications of 3D Printing: The Ultimate Guide. Available: https://amfg.ai/industrial-applications-of-3d-printing-the-ultimate-guide/#tab-con-1. [Accessed: 09-Dec-2021].

[164] 3D Printing Technology: Advantages and Disadvantages. Available: https://www.analyticssteps.com/blogs/3d-printing-technology-advantages-and-disadvantages. [Accessed: 09-Dec-2021].

[165] A.V. Golunov, S.A. Shcheglov, A.S. Golunova, O.A. Kolozova, O.V. Trapeznikova, Control and testing of the additive systems color gamut in engineering, J. Phys. Conf. Ser. 1546 (1) (May, 2020) 012062, https://doi.org/10.1088/1742-6596/1546/1/012062.

[166] A. Mahmood, et al., Applications of 3D Printing for the Advancement of Oral Dosage Forms, Elsevier Inc., 2020.

[167] Q. Meng, J. Hu, A review of shape memory polymer composites and blends, Compos. Part A Appl. Sci. Manuf. 40 (11) (November, 2009) 1661–1672, https://doi.org/10.1016/J.COMPOSITESA.2009.08.011.

CHAPTER

2

Fabrication of microchannel for water treatment using 3D printing

Prakash Bobde[1], *Ravi Kumar Patel*[2], and *Lalit Nagapurkar*[2]

[1] **University of Petroleum and Energy Studies, Dehradun, Uttarakhand, India;**
[2] **AIC-Prestige Inspire Foundation, Indore, Madhya Pradesh, India**

2.1 General

Additive manufacturing follows the following fabrication process of a required model (Fig. 2.1). The authors investigated a method for particle filtration of sizes ranging from micro to nano using shear-modulated inertial migration in straight microchannels having a rectangular cross-section [2]. The Fused Deposition Modelling (FDM) technique was used for the fabrication of microchannels. These microchannels of different structures and sizes (circular, straight, sinusoidal, 3D curved, and cross-linked curved microchannels) were fabricated by controllably extruding sacrificial molds [3]. ProJet 5500X inkjet printer was used by researchers, which utilizes a photopolymer composite material for the fabrication of robust and bendable microchannel-based strain sensor [4]. The

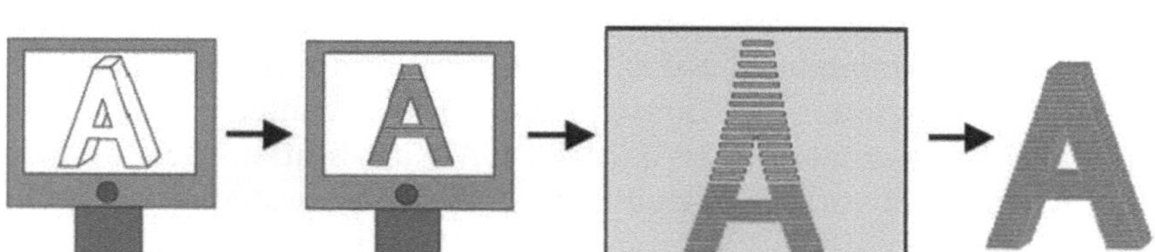

FIGURE 2.1 Additive manufacturing process [1].

https://doi.org/10.1016/B978-0-323-99861-1.00008-4

application of 3D printed microchannel has also been explored by various researchers in the detection of human breast cancer cells [5], boiling heat transfer enhancement [6], endochondral bone repair [7], and many more.

The concept of printing microchannel using MiiCraft 3D was adopted from published research work where a microfluidic channel having 1 mm diameter was printed for evaluating the milk adulteration property [8].

2.2 3D printing of microchannel

Three-dimensional printing encompasses a wide spectrum of techniques, some of which are even being used in industries. But for microfabrication, one particular technique namely "stereolithography" (SLA) is extensively used [9]. This is because some of the SLA-based printers can go up to a resolution of 56 microns like the MiiCraft 3D printer. In order to fabricate a 3D structure with a stereolithography technique, the first step is to create a digital model of the structure using computer-aided design software (CAD). The surface geometry of the 3D model has to be fabricated in the form of interconnecting tessellated triangles, followed by utilizing the dedicated software into the sequential layer of desired thickness, resolution, and accuracy. The resulting data is then transferred to the printing apparatus, which fabricates the 3D structure layer by layer starting from the bottom to top. The resulting 3D model is then exposed to UV rays for 600 s for post-curing in a different chamber. The postcuring process enhances the mechanical properties of the printed 3D structure. Fig. 2.2 depicts the procedure for the fabrication of the 3D model via stereolithography technique.

This method provides us with the flexibility of printing complex three-dimensional structures using polymeric materials [10].

2.3 Solid modeling

Solid modeling (SM) deals with CAD for the creation of geometrical shapes, surfaces, bodies, using digital toolset available in a computer graphics environment. The feature of computer graphics is to create a graphical representation of a desired physical object or part. The most known SM methods are Constructive Solid Geometry (CSG) and Boundary Representation (Brep). Many CAD applications are available out of which Solidworks 2013 published by Dassault Systèmes is used for SM.

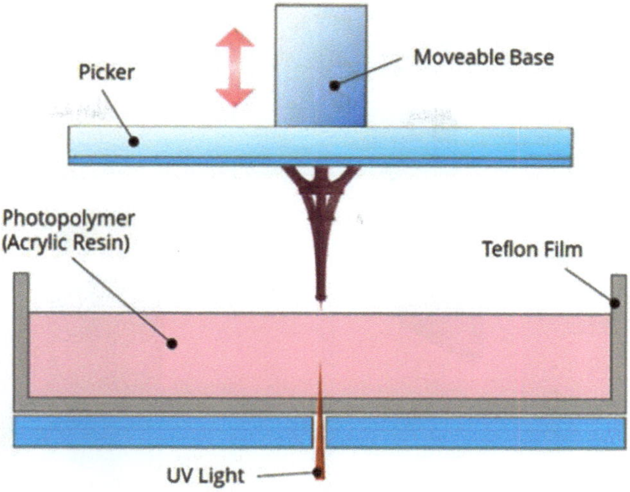

FIGURE 2.2 Stereolithography printing process.

2.4 Device design

The model of the device was designed in two categories—cuboid and cylindrical structure. In cylindrical structure two models were designed—plain microchannel and spiraled microchannel.

2.4.1 Cuboid structure

The first cuboid structure has the length, breadth, and height of 25 × 25 × 5 mm based on the maximum printable area of the 3D printer. The device consists of seven circular microchannels each having a pore size of 2 mm. The drawing and CAD model of the device is shown in Fig. 2.3A and 3D printed output of the design is shown in Fig. 2.3B. The designing of this model was done to test the proper fabrication of the channel and complete the device using a 3D printer. The reason to keep seven channels was to keep a sidewall of 3 mm thickness on both corners and 1 mm gap between each channel. The 3D printed device was tested for the channel's uniformity. This test was performed using a 5 mL syringe.

The syringe was filled with water and then from each channel, a water flow test was performed. It was found that each channel was accurately fabricated and there was no blockage. In reference to this tested model the design was further modified where the number of channels were

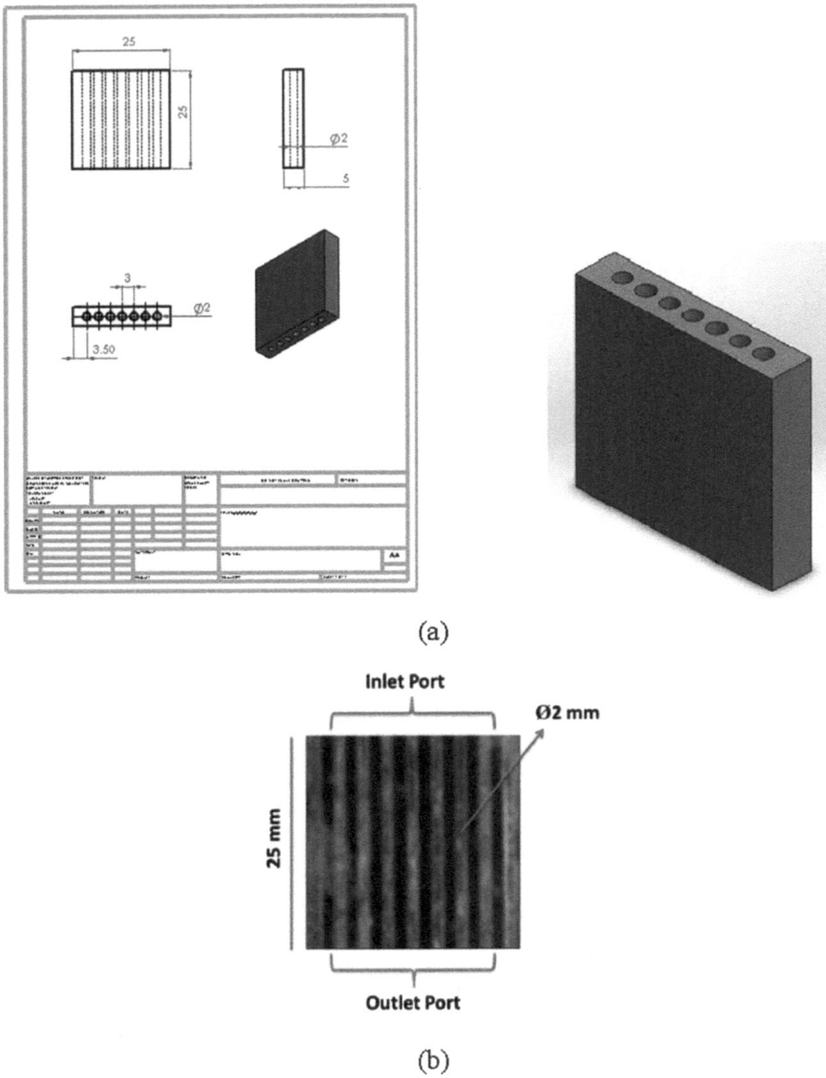

FIGURE 2.3 (A) Drawing and CAD model (isometric view) (B) 3D printed device.

increased along with the change in device dimension. Fig. 2.4 shows the drawing of the modified design.

In this modified design two separate devices were designed having a pore size of 2 mm and 1.5 mm with the same cuboid pattern having dimension 25 × 25 × 25 mm. The CAD models of devices are shown in Fig. 2.5A the top view of 2 mm pore size design and (B) the cross-sectional view of 1.5 mm pore size design.

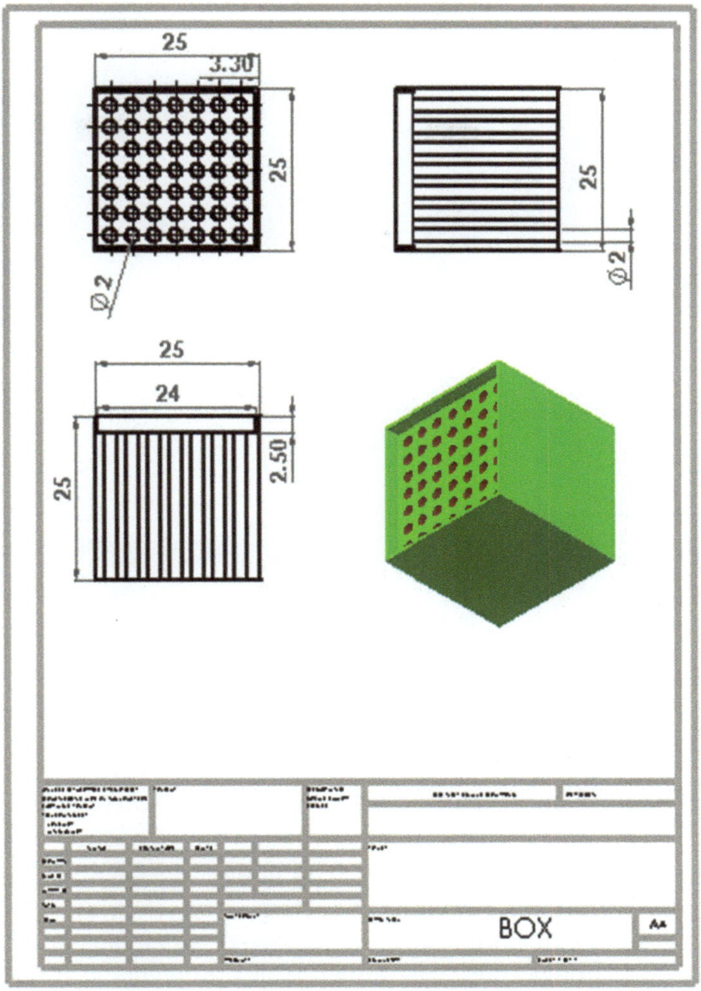

FIGURE 2.4 Drawing of the cube structure.

The 3D printed output of these designs having a 2 mm pore size and 1.5 mm pore size are shown in Fig. 2.6A and B, respectively.

These printed devices were tested for their channel uniformity by water flow test using a syringe. During the testing process for 2 mm pore size design the maximum number of channels showed uniformity in fabrication but few channels were blocked. In the same pattern the 1.5 mm pore size design was tested, and it was observed that the maximum number of channels were showing blockage. The reason for this challenge was due to an increase (congestion) in the number of

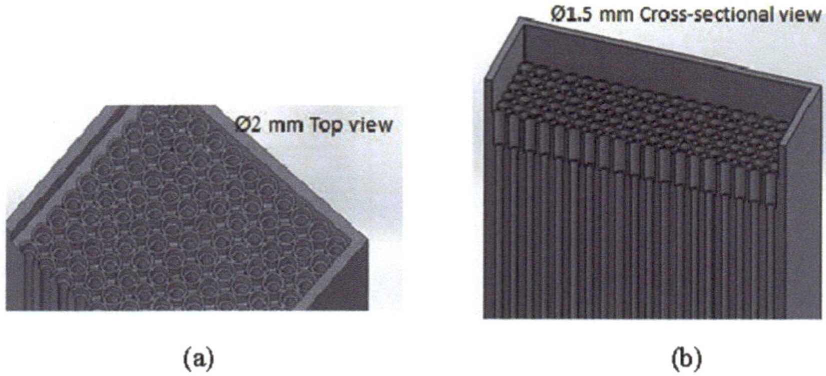

FIGURE 2.5 CAD models of the cuboid design (A) 2 mm and (B) 1.5 mm pore size.

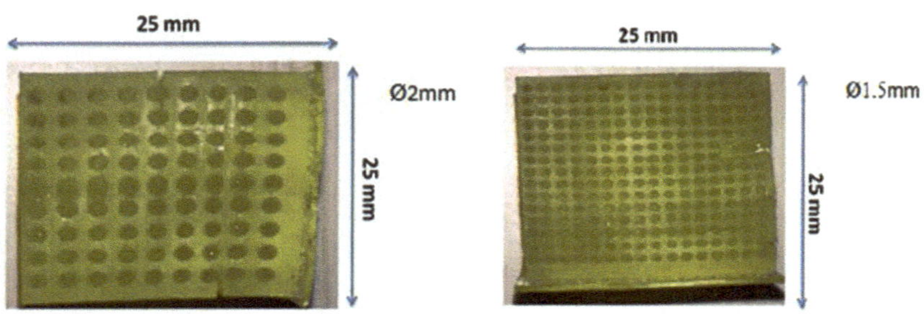

FIGURE 2.6 3D printed (A) 2 mm and (B) 1.5 mm pore size model.

channels in comparison to the previous design and also the improper resin curing during the formation of each slice.

Overall the 2 mm pore size design showed better fabrication of the device. But this device also has shown major problems in fitting to the available setup of column in our lab, as it is found that the column is generally cylindrical in shape. Due to the cylindrical pattern of the column, airtight fitting of the printed device was improper. Both the reasons such as channel blockage and improper fitting of the device led to a further modification in the design.

2.4.2 Cylindrical structure

The cuboid design and 3D printed model of cuboid have shown that the channel having a pore size of 2 and 1.5 mm can be fabricated. But there is a need for a decrease in the number of channels and the outer structure of the device. The modified design was further designed in reference to the available traditional column prototype [11,12] (Fig. 2.7).

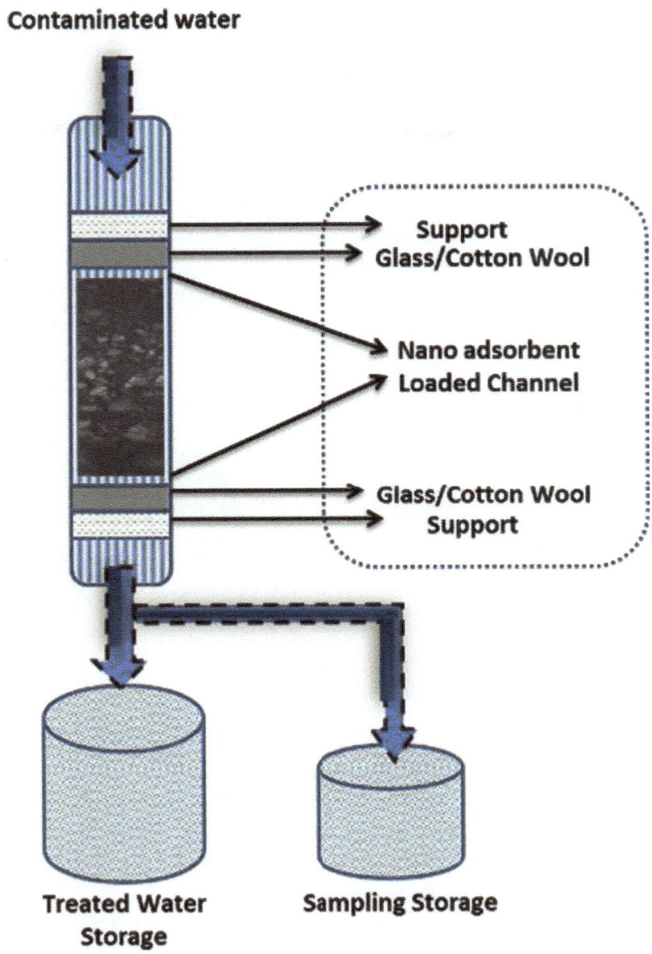

FIGURE 2.7 Tradition column prototype.

The design was divided into three parts: the inlet section, the cylindrical structure of the device, and the outlet section. The inlet and outlet sections are designed in a truncated cone shape (Fig. 2.8), the dimensions shown in the figure are in millimeters. Here the total length of the cone is 20 mm where the top surface is of 15 mm diameter which is having a pore size of 1 mm and the bottom end is kept open having a diameter of 25 mm.

The cylindrical structure of the device as shown in Fig. 2.9 was designed. The length of the device is 25 mm. Here the channels were designed with two separate models having different pore sizes of 2 mm and 1.5 mm.

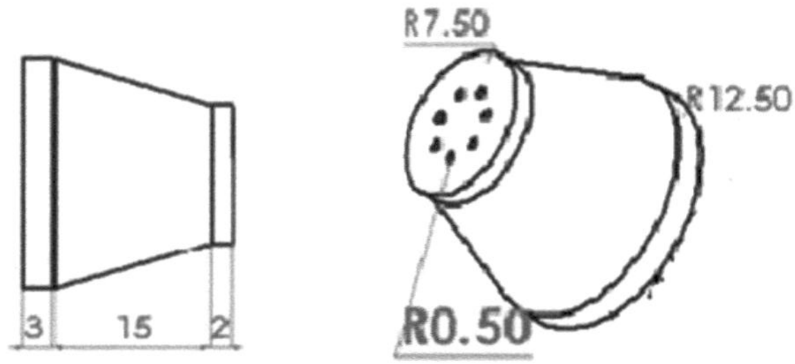

FIGURE 2.8 Inlet and outlet sections.

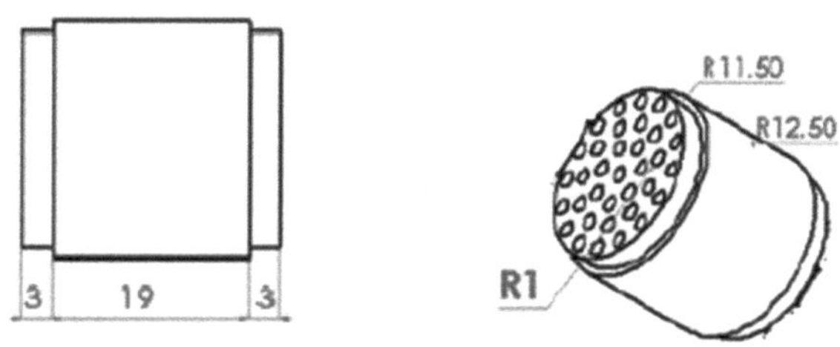

FIGURE 2.9 Cylindrical structure of the device.

The structure consists of a groove that was appropriate to assemble the truncated cone of both sides of the cylinder (Fig. 2.10).

The complete drawing and CAD model of the new design having a pore size of 2 mm are shown in Fig. 2.11A and B, respectively.

The drawing and a CAD model of design having a pore size of 1.5 mm is shown in Fig. 2.12A and B, respectively.

Such above prototypes were further implemented for fabrication using the MiiCraft 3D printer. For printing of cuboid structure the major concern was to check the proper fabrication of channels as the outer frame was previously fabricated and it was successfully printed [13].

FIGURE 2.10 Final assembled CAD model.

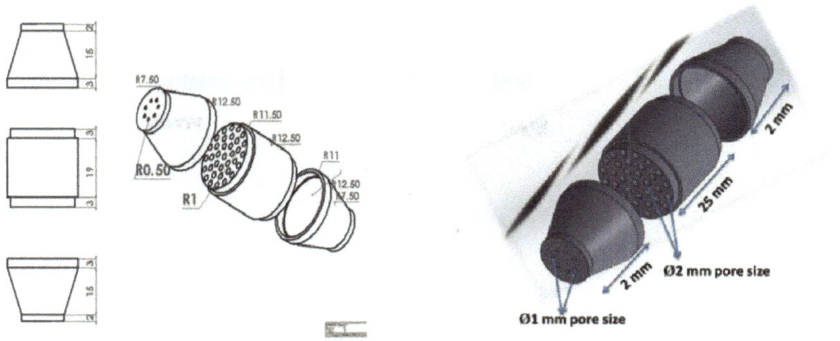

FIGURE 2.11 (A) Drawing and (B) CAD model of cylindrical design.

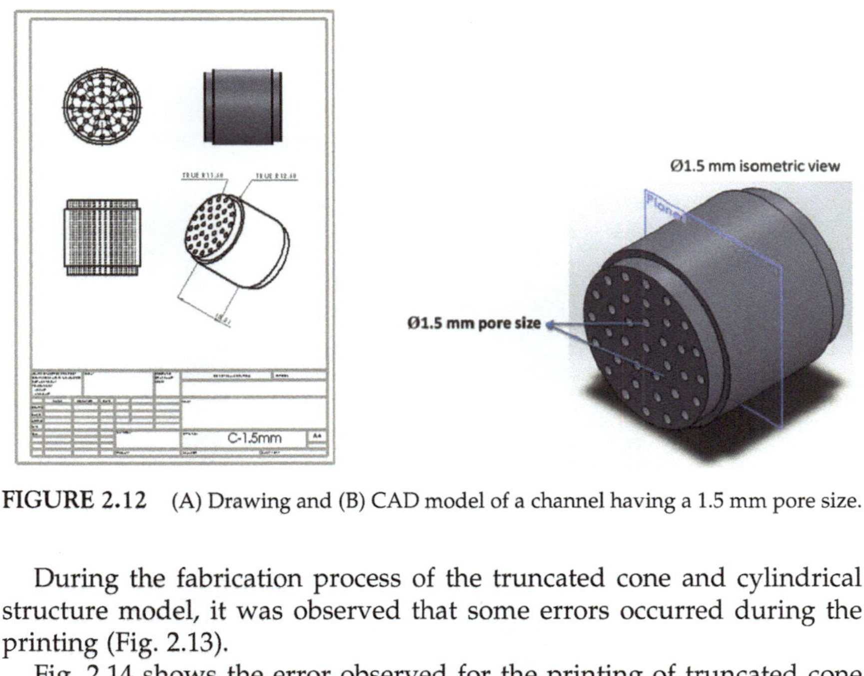

FIGURE 2.12 (A) Drawing and (B) CAD model of a channel having a 1.5 mm pore size.

During the fabrication process of the truncated cone and cylindrical structure model, it was observed that some errors occurred during the printing (Fig. 2.13).

Fig. 2.14 shows the error observed for the printing of truncated cone where the shape of the cone got deformed due to the elongation of the corner edges (Fig. 2.14A) and the incomplete printing of the cone (Fig. 2.14B).

But for the printing of cylindrical structure (Fig. 2.15) it was found that the last layer of the device was not formed properly.

The above observed printing errors occurred due to the printing operation setting, resin tank, and the printer nozzle maintenance (Fig. 2.16). The printing operation here was varied in accordance with the

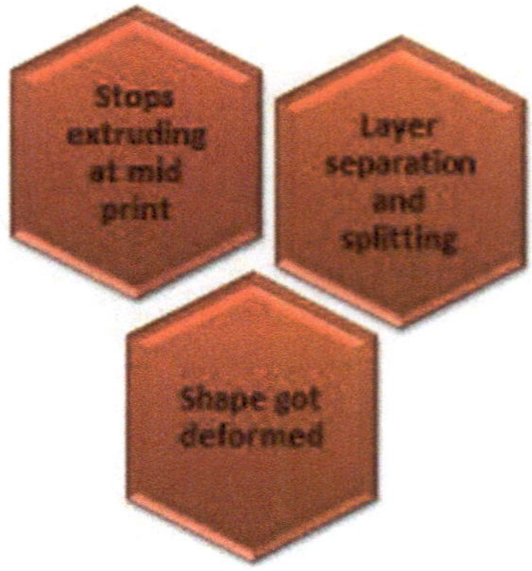

FIGURE 2.13 Printing errors.

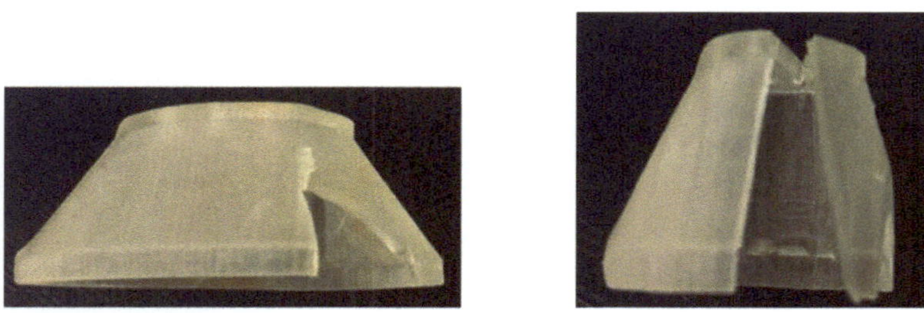

(a) (b)

FIGURE 2.14 Truncated cone (A) Elongation of cone (B) Splitting edge.

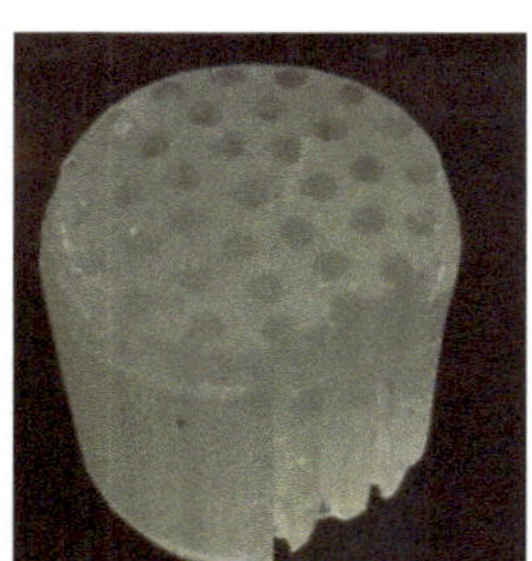

FIGURE 2.15 Cylindrical structure deformation.

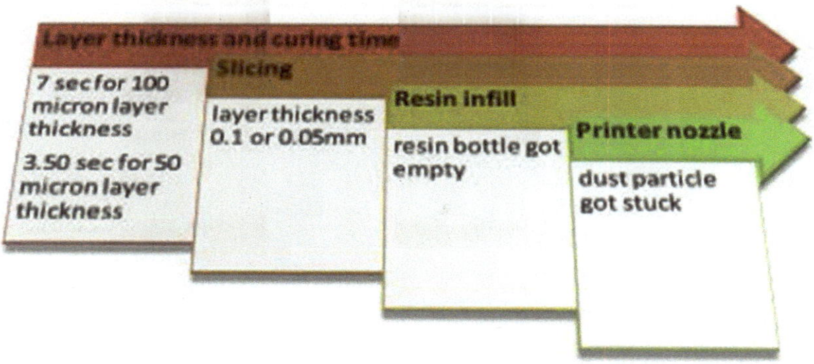

FIGURE 2.16 Printing error parameters.

printing of the cuboid structure as this design was attempted for the first time with this printer in our R&D lab.

All the errors were considered that occurred during the initial printing process of the designed model. The printer setting was done as per requirement for the proper fabrication of the truncated cone and cylindrical structure. The perfectly 3D printed models of the designed device parts are shown in Fig. 2.17.

The two different cylindrical structures having 2 mm pore size channels and 1.5 mm pore size channels were printed accordingly using the MiiCraft 3D printer as shown in Fig. 2.18. Both the structures, 2 mm pore size and 1.5 mm pore size, have a total number of 35 channels and the pattern the channels were circular and plain in design.

These fabricated channels were checked for uniformity by performing a water flow test where again a syringe was used. The syringe was filled with water and from each channel water was passed to check the

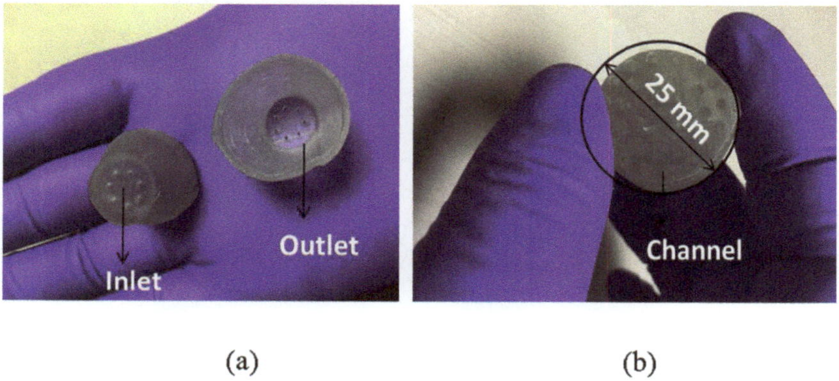

FIGURE 2.17 (A) Inlet and outlet section (B) Cylindrical structure.

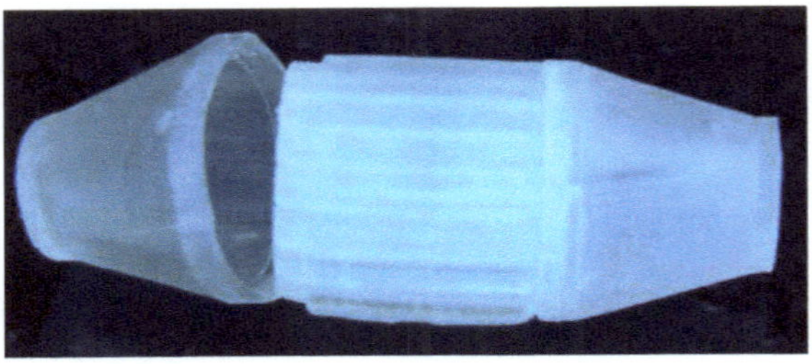

FIGURE 2.18 3D printed parts of the filter.

continuity. In comparison to the cuboid structure the number of channels was reduced to a large extent and there was proper symmetry between each channel. This change in design support positively during the fabrication of the device and the test results also supported it. In the process of water flow test from the channels it was found that the channels were uniformly fabricated and there was no occurrence of blockage.

2.4.2.1 *Spiraled microchannel*

In accordance with the successful fabrication of cylindrical structure having plain microchannel, the design was further modified. In modified design the structural design used was cylindrical having a channel of 2 mm pore size. The modification in the design is shown in the drawing sheet (Fig. 2.19) where the internal surface of the channels is now designed in the spiral form.

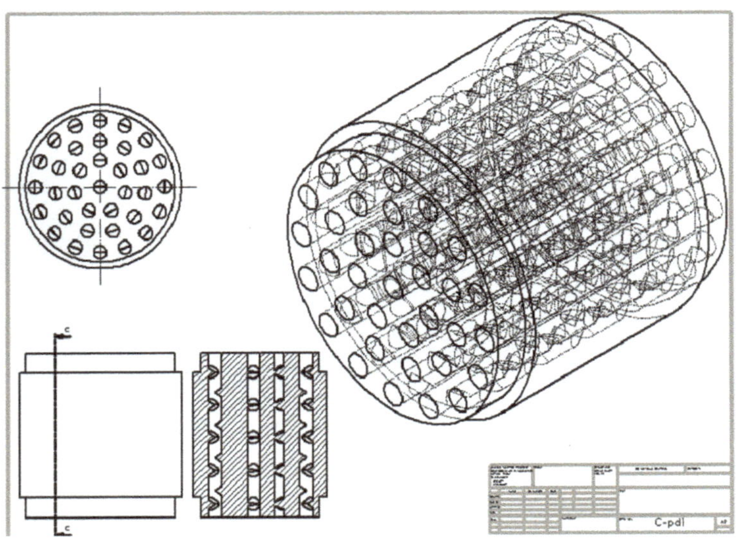

FIGURE 2.19 Drawing of spiraled microchannels.

Fig. 2.20 shows the cross-section view of the CAD model of the modified design for the cylindrical structure.

This new design was further fabricated using a 3D printer and tested for the proper formation of channels. The steps involved in testing were the same as was for the plain circular channel. In this new model the outlet section and the inlet section are the same as per previous designs. After the testing procedure it was observed that the spiral design was successfully fabricated.

2.5 TMA analyzer

The performed TA Q-400EM is helpful to acquire the Tg from their temperature modified dimension changes of the truncated cone layer. In TA-Q400EM, a three-point bending test was performed to address the storage modulus, loss modulus, stress relaxation, and stress-strain behavior of the rectangular specimens. All dimensional change, storage modulus, and loss modulus experiments were performed for the temperature range of 40°C using heating rates of 10°C/min with 50 mL/min nitrogen purge gas flow and 0.02 N force. In the stress relaxation study, 1×10^{-3} N preloaded force was applied to straighten the specimen. During testing, the required temperature and 1% strain were applied, then relaxation modulus was evaluated with respect to time. Stress-strain experiments were performed in strain ramp mode with 0.02 N force and strain was evaluated at 40°C isothermal temperature.

The stress-strain analysis performed showed that the fabricated truncated cone is having a flexible property at a temperature of 40°C (Fig. 2.21).

This flexibility shows that the part is durable and can be used for the required purpose.

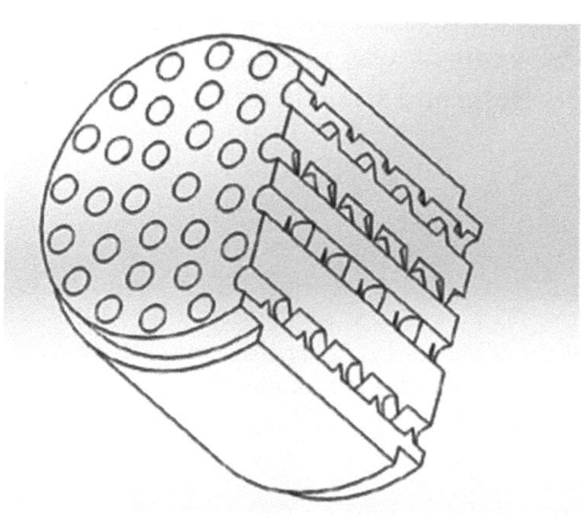

FIGURE 2.20 Cross-sectional CAD view showing the internal pattern of channel.

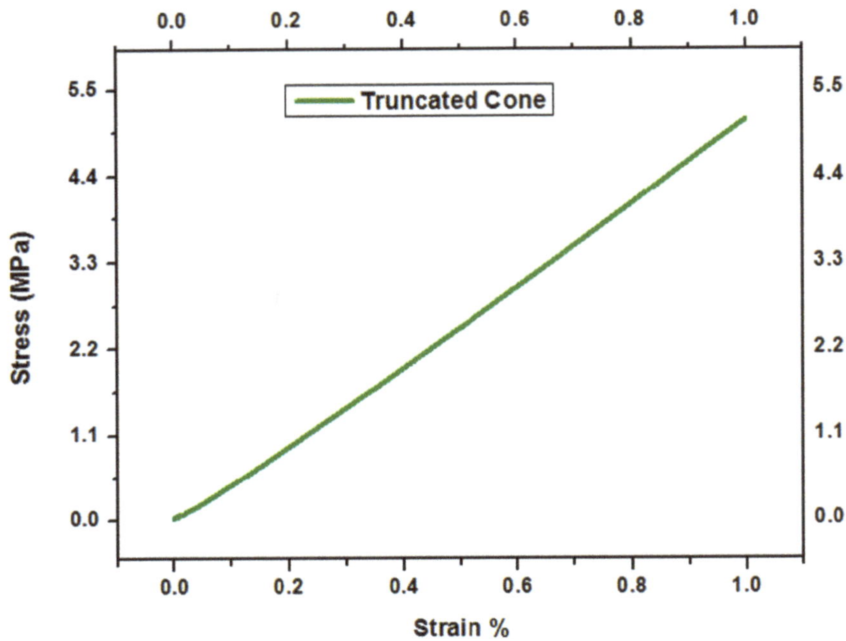

FIGURE 2.21 Stress and strain analysis response.

2.6 Conclusion

The development of the device with reference to the available traditional column setup was a major concern. The purpose of designing a device was to use and implement 3D printing technology for water treatment applications. Here various design layouts were prepared using Solidworks software and then fabricated using the MiiCraft 3D printer. These printed devices were tested for their channels' continuity formation and their adaptability. After all testing and finalization procedures, cylindrical structure and truncated cone were observed to better fit with the available filtration setup. The TMA analysis was also performed on the layer of the truncated cone to test its stress and strain bearing capacity and its response was appreciable.

References

[1] Phys.org, Does 3D Printing Have the Right Stuff?, 2014. Retrieved from, https://phys.org/news/2014-06-3d.html.

[2] A.A.S. Bhagat, S.S. Kuntaegowdanahalli, I. Papautsky, Enhanced particle filtration in straight microchannels using shear-modulated inertial migration, Phys. Fluids 20 (10) (2008) 101702.

[3] W. Tang, H. Liu, L. Zhu, J. Shi, Z. Li, N. Xiang, J. Yang, Fabrication of different microchannels by adjusting the extrusion parameters for sacrificial molds, Micromachines 10 (8) (2019) 544.
[4] S. Agarwala, G.L. Goh, Y.L. Yap, G.D. Goh, H. Yu, W.Y. Yeong, T. Tran, Development of bendable strain sensor with embedded microchannels using 3D printing, Sensor Actuator Phys. 263 (2017) 593–599.
[5] H. Motaghi, S. Ziyaee, M.A. Mehrgardi, A.A. Kajani, A.-K. Bordbar, Electrochemiluminescence detection of human breast cancer cells using aptamer modified bipolar electrode mounted into 3D printed microchannel, Biosens. Bioelectron. 118 (2018) 217–223.
[6] W. Zhang, Y. Chai, J. Xu, G. Liu, Y. Sun, 3D heterogeneous wetting microchannel surfaces for boiling heat transfer enhancement, Appl. Surf. Sci. 457 (2018) 891–901.
[7] A.C. Daly, P. Pitacco, J. Nulty, G.M. Cunniffe, D.J. Kelly, 3D printed microchannel networks to direct vascularisation during endochondral bone repair, Biomaterials 162 (2018) 34–46.
[8] P.S. Venkateswaran, A. Sharma, S. Dubey, A. Agarwal, S. Goel, Rapid and automated measurement of milk adulteration using a 3D printed optofluidic microviscometer (OMV), IEEE Sensor. J. 16 (9) (2016) 3000–3007.
[9] T.J. Snyder, M. Andrews, M. Weislogel, P. Moeck, J. Stone-Sundberg, D. Birkes, O. Fercak, 3D systems' technology overview and new applications in manufacturing, engineering, science, and education, 3D Print. Addit. Manuf. 1 (3) (2014) 169–176.
[10] P.F. O'Neill, A. Ben Azouz, M. Vazquez, J. Liu, S. Marczak, Z. Slouka, D. Brabazon, Advances in three-dimensional rapid prototyping of microfluidic devices for biological applications, Biomicrofluidics 8 (5) (2014) 052112.
[11] J. García-Sánchez, M. Solache-Ríos, V. Martínez-Miranda, R. Enciso-Perez, N. Arteaga-Larios, M. Ojeda-Escamilla, I. Rodríguez-Torres, Experimental study of the adsorption of fluoride by modified magnetite using a continuous flow system and numerical simulation, Process Saf. Environ. Protect. 109 (2017) 130–139.
[12] S. Mohan, D.K. Singh, V. Kumar, S.H. Hasan, Effective removal of Fluoride ions by rGO/ZrO2 nanocomposite from aqueous solution: fixed bed column adsorption modelling and its adsorption mechanism, J. Fluor. Chem. 194 (2017) 40–50.
[13] R.K. Patel, M.K. Awasthi, Experimental analysis of 3D printed microfluidic device for detection of adulteration in fluids, in: Paper Presented at the International Conference on Nano for Energy and Water, 2017.

CHAPTER

3

Recent advances on 3D printing for wastewater treatment and process optimization using artificial intelligence and machine learning: updates and perspectives

S. Tripathy[1], *D.K. Tripathy*[2], *and S. Samantaray*[1]

[1] **Department of Mechanical Engineering, ITER, Siksha 'O' Anusandhan (Deemed to be University), Bhubaneswar, Odisha, India;** [2] **IIT Kharagpur, Kharagpur, West Bengal, India**

3.1 Introduction

3D printing technology with its high manufacturing flexibility has become one of the most progressive methods for various engineering and nonengineering applications like construction, aerospace, biomedical, food industries, regenerative medicine, diagnosis, implants, artificial tissues, organs, prosthetics, cancer research, etc. The technique uses layer-by-layer deposition of material to fabricate objects with complex geometrical constraints from 3D model data by adjusting the shape of each layer. The method is extremely advantageous over other conventional manufacturing methods in terms of high reproducibility and control, speediness, cost-effectiveness, customization with no restrictions on the spatial arrangement, flexibility, accessibility, and sustainability with reduced risk. It can be used for a wide range of materials like polymers, ceramics, metals, composites, etc. [1]. The limitations of the process are that it cannot be employed for mass production for the complexity in printing

3D Printing Technology for Water Treatment Applications
https://doi.org/10.1016/B978-0-323-99861-1.00003-5

methods and associated cost. The process is not suitable for manufacturing large structures due to likely printing defects and the challenges in selection of compatible materials [2]. The major goal of sustainable manufacturing is to minimize the negative environmental impacts and conserve the natural resources for future generations. For the sustainability of water and energy systems, modeling the interactions, purification, and recycling of the resources have become imperative. Needs of quality water for living beings, improved living standard of people coupled with availability of various water purification techniques has led to increased global demand for high-quality drinking and water for other uses. The characteristic of wastewater due to the presence of different pollutants in varied proportions makes the treatment process complex. 3D printing techniques find significant importance in fabrication of membranes and other complex water treatment components. Application of 3D printing in wastewater treatment and water environment quality management to achieve sustainable solutions for energy recycling has gained focus of researchers in a study flagging the challenges in the process [3]. The application of data-driven methods, such as artificial intelligence (AI) and machine learning (ML), would help in overcoming these challenges. The AI and ML techniques have found a wide range of applications in research and commercial fields recently and have been proved to be successful with performance more than human level [4]. AI and ML in combination with big data are a new area of interest for researchers and have been focused in review articles and technical articles in areas of mathematics, statistics, computer science, energy engineering, medicines, etc. The data-driven techniques are advantageous as they do not need any physical description of the processes to acquire the data for establishing the relationship between variables in the system. The scientific processes and the underlying principles make these techniques superior and useful over other methods. However, possibilities of miscalibration or parameter instability stand as limitations of these techniques. Sufficient data must be available from the system, in absence of which the analysis will be time consuming and may lead to uncertain outcome. Complete knowledge on the system and interaction of resources will provide better predictive results. The water data collected by agencies through surveys and reports may be used, and the data-driven techniques can be effectively applied to understand and predict the interaction of resources within the wastewater system [5].

3.2 Overview and scope

3D printing technology with its promising potential is being applied as a prospective method for water treatment. It has been successfully applied for the development of module spacers, filtration, desalination membranes, water remediation, etc. The technology has offered high printing resolution, better efficiency, and faster production rates, which have attracted research interests in the domain. The techniques of AI and ML have been implemented for wastewater treatment in the past, but its application to water treatment using 3D printing technology needs to be investigated further. This review article is an effort in this direction and will be a resource of research on 3D printing for water treatment technology using AI and ML. The review focuses on the advantages of using 3D printing for wastewater management, purification methods using 3D-printed membranes, artificial intelligence, and machine learning applications in 3D printing techniques for wastewater treatment for the benefit of researchers and readers. This article includes the following sections:

- Section 1 defines scope of the literature survey including sources and keywords. It also includes the Bibliometric and review methods in the field of wastewater treatment. The research trends and updates of the water treatment technology have been highlighted along with the various methods applied for water treatment.
- Section 2 describes the principles and techniques of membrane separation technology and membrane modules required for the wastewater treatment process.
- Section 3 highlights the current applications of 3D printing technology for fabrication of module spacers and membrane filters. The emerging trends of 3D printing methods have been portrayed. The advanced applications of 3D printing for water treatment have also been discussed in detail. The computational techniques for cost optimization and sustainability have been described. The traditional methods have been compared with the superhuman methods. The challenges of data-driven techniques like artificial intelligence and machine learning for carrying out the analysis of wastewater treatment have been discussed while keeping in view the previous research work. It provides an overview on the use of 3D printing technology for wastewater treatment using data-driven methods for cost and time optimization. The insights on data-based modeling, fault detection, and diagnosis with proper analysis have been provided and relevant future research directions using AI and ML applications for 3D-printed wastewater technology have been discussed.
- Section 4 summarizes the possibilities of developments in the process and suggests improvements that would help the engineers to determine the methods and processes and identify the most promising technique to overcome challenges during execution.

3.2.1 Bibliometric and review methods

There has been tremendous development in the field of 3D printing technology during last one decade, particularly in the field of water purification and membrane separation. In 2010, there were nearly 20 articles focusing on 3D printing and its application, increased to about 300 in last 5 years. The majority of the study focused on applications and scope of 3D printing, while others on its efficiency specifically for making membrane spacers for wastewater treatment. A recent review article has highlighted the developments in design, fabrication methods, and performance while using 3D-printed materials for membrane separation, water purification, desalination, and water treatment. With the tremendous growth and success of data-driven applications, its use in 3D-printed water purification methods needs to be further investigated. To provide structured information on application of 3D printing for wastewater treatment, relevant to artificial intelligence, and machine learning methods, the present article focuses on scope of 3D printing for water treatment. The present study does not intend to cover the entire field of AI and ML research and scope for future work in the field of data analytics. It is limited to application of 3D printing for wastewater treatment with use of AI and ML methods. Bibliometric search is an important way to explore and analyze the research trends. Relevant documents have been searched with keywords, co-words, titles, and abstracts with terms "3D printing for wastewater treatment" and "artificial intelligence and machine learning for wastewater treatment" published in SCI indexed and other published articles. Based on this analysis, a detailed review has been conducted on 3D printing applications in wastewater treatment, preparation of membranes and purification, technological performance, economic cost analysis, management, and wastewater recycling. Some thoughts in the future research directions have also been presented. The summaries searched from various published articles are presented as a resource to the wastewater treatment technology in order to aid research and application of 3D printing in the relevant field. Applications of AI and ML techniques have also been focused to enhance the efficiency of the process of water treatment and for cost optimization.

3.2.2 Research trends in wastewater treatment

Water quality management has been one of most crucial problems faced by the world in the recent years as it affects all living beings. Water quality depends on its physical, chemical, and biological properties. The key variables in the wastewater treatment need to be assessed to control the water pollution. Mostly, the diseases faced by living beings are water borne and to protect human health and natural ecosystem; the water environment quality must be maintained. The characteristic of wastewater is very complex due to presence of different pollutants distributed in varied proportions. The causes of water contamination are domestic, industrial, infiltration, and mines. Water is classified into four basic types: (i) rainwater, (ii) domestic wastewater, (iii) agricultural wastewater, and (iv) industrial wastewater. The large-scale use of water in industries like petrochemicals, refinery and fuel production plants, agriculture and food processing, textile, and leather industries make it potentially most toxic. Wastewater treatment is thus a very complex process, which depends on the uncertainty and proportion, quantity, quality of influent, and emission efficiency of the wastewater [6]. The water purification process comprises of five steps: (i) pretreatment (physical and mechanical), (ii) primary treatment (physiochemical and chemical), (iii) secondary treatment (chemical and biological), (iv) tertiary treatment (physical and chemical), and (v) treatment of the sludge formed (recycling). Multiple techniques are used for the wastewater purification and reuse. The water quality management method has evolved from the traditional laboratory based on in-situ methods. The traditional methods for water quality inspection include physical observation where the consumers offer insights on the safety of the use of water based upon specific water quality indicators. Laboratory-based analysis is used where samples are taken at specific time intervals, and equipment is used to detect the contaminants present. This method predicts the water quality for drinking and other purposes. Wireless portable devices are also used for contaminant detection in water bodies. The traditional methods employed do not provide accurate results and are unable to predict the contaminant present in low concentration, low volatility, different hydrophobicity, variants of microorganisms, interaction of solutes, and medium of separation. These methods are not suitable for real-time measurement and evaluate the heterogeneous parameters present in water bodies [7,8]. Soft sensor techniques are a good solution to such problems, and they can measure the variables by correlating them with the available data of variables. Model-driven and data-driven techniques are such soft sensor methods. Model-driven techniques are difficult to be implemented as modeling of the entire process at a preliminary stage with insufficient information is difficult. The data-driven models are based on historical data patterns obtained from industrial processes without any prior knowledge and are acceptable to model complex processes [9,10]. The feasibility, cost, reliability, environmental impact, difficulty of operation, and the wastewater quality determine the technique of purification to be used. The limitations of traditional methods led to the development of membrane separation, which has many advantages like high efficiency to remove contaminants, organic matter, salts, and metals. The membrane separation technology has observed significant application in the past due to its low capital, size of equipment, and energy requirement. The efficiency of membranes need to be improved further as they concentrate the pollutant but do not degrade it. The increase in the concentration of the pollutant decreases the filtration capacity of the membranes reducing its effectiveness. The potential of the membrane separation process has been investigated, and improvements have been proposed by researchers in the previous studies. Much focus on membrane technology has been given to microfiltration, nanofiltration, ultrafiltration, reverse osmosis, electrodialysis, capacitive deionization, membrane distillation, membrane bioreactors, etc. These processes are basically pressure driven where the difference in pressure acts as the driving force. The driving force for movement of media through the membranes may be in equilibrium or nonequilibrium, and they can be pressure-driven or non-pressure-driven. The techniques based on their driving forces may be broadly classified as follows:

1. Non-equilibrium-based and pressure-driven: Microfiltration, ultrafiltration, nanofiltration, reverse osmosis, and pervaporation
2. Non-equilibrium-based, non-pressure-, electrically driven: Electrodialysis
3. Equilibrium-based process that is pressure- and thermal-driven: Membrane distillation
4. Equilibrium-based and non-pressure-driven: Forward osmosis and liquid gap membrane distillation.

3.3 Principles

The hydraulic pressure difference when used as the driving force for mass transfer has many disadvantages. For desalination during reverse osmosis or nanofiltration processes, osmotic pressure limitation occurs [11].

An overview on the principle of membrane separation process has been presented in this section.

3.3.1 Membrane separation technology

Membrane separation is one of the most important methods to produce good quality water with higher removal rates. Depending on the structure and composition, membranes may be isotropic or anisotropic. They can also be microporous where the fluxes are high or nonporous where the fluxes are low. Microfiltration, ultrafiltration, and microbioreactors are used to purify raw contaminants, whereas nanofiltration, reverse osmosis, and membrane distillation are advanced treatments used for polishing. The technique for removal of flux involves three mechanisms: size exclusion, electrostatic repulsion, and adsorption. The membrane bioreactors use aerobic, anoxic, and anaerobic biological treatment methods along with microfiltration or ultrafiltration. The method has been successful for the removal of micropollutants, due to integration of two separate processes. Nanofiltration and reverse osmosis have been proved to be very effective techniques for removal of pollutants with the limitation of higher energy consumption. The membrane being denser demonstrates greater efficiency and has gained much research interest in the past few years. To get improvised results using reverse osmosis process, different membrane materials have been used [12]. Membrane materials are usually of two types: organic and inorganic. Organic membranes are made up of synthetic polymers, which include polyethylene, polypropylene, polytetrafloroethylene, and cellulose acetate. The inorganic membrane materials are ceramics, zeolites, metals, or silica, which being thermally and chemically stable find increased industrial application for nanofiltration, ultrafiltration, and hydrogen separation [13, 14]. Results show that polyamide membranes provide good rejection in comparison to membranes made up of cellulose acetate as electrostatic exclusion is an important phenomenon taking place during the process. Pervaporation combines permeability and evaporation for the separation of liquids where the feeding of the liquid occurs on one side evaporating the permeable component from the other side. Electrodialysis and electrodialysis reversal are used to separate dissolved ions from water by combining electricity and ion permeable membranes. In such processes, two forms of ion exchange membranes are used. The one which rejects cations is permeable to anions, whereas the other one is permeable to cations and rejects anions. Two streams of solutions are passed, i.e., the concentrate and the feed. Under the influence of electric current, ions move from feed to concentrate through oppositely charged membranes for anion exchange and cation exchange. The process is not appropriate for high saline water as the ions removed and desalination energy is proportional. Membrane distillation operates by the transportation of water in vapor state when passed by a membrane, which is microporous and hydrophobic. Total retention of nonvolatile components has been observed, and a difference in temperature between the feed and the product is noticed [15]. The water gets vaporized at the pores of the feed and gets condensed at the product membrane. The membrane distillation is advantageous as it can be combined with a renewable energy source, such as solar, wind, and geothermal along with waste thermal energy. In liquid gap membrane distillation, a nonpermeable foil is used as a channel to separate the permeable component from the coolant [16,17].

3.3.2 Membrane modules

Membrane separation processes employed in industries require large membrane areas and their economical packing are referred to as modules. These modules are classified into four major types viz. tubular, plate and frame, spiral wound, and hollow fiber modules. Tubular module consists of a tubular shell, which is a perforated stainless steel or fiber glass pipe embedded within by a semi-permeable membrane, used for feed with high solid content. Plate and frame module consists of a membrane, feed spacer and product spacer embedded together in a metallic frame. The spacers are used as channels for the feed and product flow while preventing sticking of the membranes. Spiral wound module contains membranes and spacers coiled around a perforated central collection tube, which is placed in a tubular pressure vessel. Water enters tangentially to the module and the permeable component flows perpendicularly to the membranes and get deposited in the collection tube. These spacers find much application in reverse osmosis and nanofiltration processes as replacement of the module elements is easier [18]. The hollow fiber module houses a bunch of closed- or open-ended hollow fibers in a pressure vessel. These modules are either shell-side, used for high pressure applications or bore-side, used for low- to medium-pressure applications and involve high capital investment.

There has been lot of advancement in the field of wastewater treatment using membrane technology; still, there remains space for improvement as water quality is changing with time and so are the pollutants. To deal with the challenge of membrane cleaning and fouling, chemical cleaning methods are used, which require higher energy as well as cost involvement. The development of cheaper membranes with resistance to fouling will make the wastewater treatment process more viable and suitable for large scale applications.

3.4 Current applications

The improvement in the fabrication methodologies of water treatment components is a major concern of researchers. The current applications and recent developments of the process are discussed subsequently.

3.4.1 3D printing technologies in wastewater treatment

3D printing possesses a huge potential for wastewater treatment compared to other traditional methods. More than 50 techniques of 3D printing have been identified by American Society for Testing and Materials (ASTM). These are broadly classified into seven classes: ink-jet printing-assisted fabrication methods, powder bed fusion technologies, binder jet, extrusion, vat photo polymerization, sheet lamination, and direct energy deposition. However, limitation of the process is mechanical anisotropy of the compatible material. The weaker interlayer bonding between layers makes the tensile properties weak. The 3D printing technology has gained acceptance over traditional manufacturing methods due to its high precision, design flexibility, material saving, and customization possibility. Emerging methods of 3D printing and new materials compatible with the process of manufacture are continuously being developed. This technology in the meanwhile has penetrated beyond the laboratories to domestic level because of the reduced cost of 3D printers. The fabrication methodologies of water treatment components using 3D printing technology along with the modifications required to enhance the performance of membranes have been discussed subsequently.

3.4.1.1 *Ink jet printing*

The process involves pumping and deposition of droplets of a stable ceramic suspension like zirconium oxide powder in water onto the substrate through an injection nozzle. The droplets form a continuous pattern and get solidified to hold the successive layers in position. The method is fast, effective, and flexible for printing complex structures. The inks are either made up of wax or liquid suspensions.

3.4.1.2 *Powder bed fusion*

A thermal source is required for sintering, melting, and fusion of the powder particles. It is followed by rolling layers of powders on the fused ones, till the entire design is created. Selective laser sintering, selective laser melting, and electron beam melting are the examples of the process.

3.4.1.3 *Binder jet*

Liquid binder is printed using inkjet mechanism on a layer of powder spread on the bed. It is sequentially followed by spreading another layer of powder and printing more liquid binder. The binding agent helps the powder particles adhere and create the model by layer wise deposition.

3.4.1.4 *Material extrusion*

The material is forced out through a printing head due to application of a constant pressure on the printing platform. The extruded material gets solidified, and the deposition continues until the final model is created. Fused deposition modeling and direct laser writing operate under this technique.

3.4.1.5 *Vat photo polymerization*

Selective curing of a photopolymer is done using a light source in a vat. On being hit with the laser source, the resin gets solidified. A fresh layer of resin is again deposited on the solidified layer to be cured. The process is repeated until the final prototype is achieved. This technique is mostly applied for the fabrication of membranes. Vat photo polymerization includes stereo lithography, digital light processing (DLP), photopolymer jetting, and continuous liquid interface production (CLIP).

3.4.1.6 *Sheet lamination*

The sheets of printing material are placed in layers, cut in shapes, and jointly laminated forming the design proposed. The sheets are placed on the platform and heated rollers are allowed to move on it with application of pressure to make a bond between the solidified layer and the current layer. An adhesive coating is applied between the sheets for better joining. The platform is lowered for feeding a new sheet on the bed. Deposition lamination, laminated object manufacturing (LOM), and ultrasonic additive manufacturing are some of the processes under this technique.

3.4.1.7 *Direct energy deposition*

The laser beam is focused on a small region of the printing material, which starts melting after getting heated up. Industrial metal powder is used as the printing material, and the printing resolution determines the amount of powder melted [19].

3.5 Application of 3D printing for water treatment using membrane technology

Fabrication of membranes in conventional methods face challenges like misalignment of supporting transport channels and separating layers and adherence between the phases and the support. 3D printing technology is capable of supporting multitude designs and has the potential to produce low-cost membranes where microporous materials are combined with the printing material, and the final product reaches up to the size of microns. The applications of 3D printing for wastewater management have been explained hereunder.

3.5.1 Membrane module design

Membrane modules consist of membranes, inlet for feed water stream, pressure supports, outlet stream for the permeable component with a support structure to provide the necessary support system. The membranes are either of flat sheet or hollow fiber type, which are then packed into tubes to form the modules. The membrane modules used in the RO process are either of spiral wound type or hollow fiber module type and offer an optimal balance between operation, packing density, fouling control, and efficiency.

3.5.2 Design and fabrication of module spacers

To enhance the fluid mixing and improve the performance of modules, net-like feed spacers are placed between the membranes to form flow channels and keep membranes separated. Their structure requires flexibility in the design so that they can wound around the tube while maintaining the required stiffness. Polypropylene is the most widely used material due to its chemical resistance and flexibility. Simulation studies have been performed by researchers to optimize the performance of the spacers. The application of 3D printing for fabrication of spacers was done 17 years ago where the study on mass transfer enhancement between spacers was also carried out. The performance of multilayered spacers was found to be superior to twisted tape and modified filament spacers. 3D printing of spacers improves the hydrodynamics and reduces the befouling [20]. Experiments have been performed to compare 3D polyjet–printed spacers to conventional spacers. Numerical modeling and experimental testing of 3D printed spacers can enhance the performance of membranes further. Studies have shown that out of other 3D printing techniques, the polyjet technique produces spacers with better resolution and lower cost but has higher affinity toward bacteria attachment. Selective laser sintering was found to be the most suitable 3D printing technique for fabrication of spacers [3].

3.5.3 Design and fabrication of membranes

Due to the large range of resolution from 0.1 to 10 microns, 3D printing technique is less used for fabrication of membranes in comparison to fabrication of spacers. It is applied for microfiltration where large pore membranes are used like MF and its application to reverse osmosis, NF and UF will still require further studies.

3.6 Membrane materials and their fabrication methods

Mostly the membranes used are made up of polymers, but ceramic membranes also find applications. Polymeric membranes are prepared using phase inversion process, stretching, extrusion, hollow fiber spinning, etc. The 3D printing technique finds limited use for membrane fabrication due to the limitation of achieving the desired material characteristics and wettability. In composite membranes, 3D-printed layers are used as substrates and preparation of active layers uses other methods. Composite membranes produce comparative results to polyethersulfone (PES) mixed membranes and are superior to pure (PES) membranes [19]. Sintering temperature has significant effect on morphological and structural features of ceramic membranes. With the increase in sintering temperature, the pore size increases. To reduce the sintering temperature for ceramic membranes during fabrication using 3D printing methods, fine powders may be used for processing, agglomerates may be removed, and additives for sintering may be used. Thermal binding is used to remove the soluble and insoluble components of the binder. During postprocessing, the pores are filled with wax and the process is called infiltration. It is carried out to provide strength to the final part, which is otherwise porous and delicate. Ag and Cu nanopowders have been added to fresh geopolymer paste for making modified filters [21]. Metals have been found to leak out during the water removal. While using salts of Ag and Cu, the colors of the filters were changed, and metal ions were trapped inside the network. Direct foaming methods do not cause, $AgNO_3$ or Cu $(NO_3)_2$ modifications as the transition metal decompose hydrogen peroxide without forming foams. 3D printing has been used to fabricate carbon nitride–based hybrid aerogel membranes with patterned macroscopic architectures [22]. Nanosheets of g-C_3N_4 along with sodium alginate have been 3D printed using air or reservoirs of $CaCl_2$/glycerol solution or pluronic F127. The printed aerogel membranes exhibit visible light absorption, superior performance for solar wastewater remediation with remarkable cyclic stability. Another article reports the development of a multiuse membrane using 3D printing technology by Nano Sun, a university-based start-up, which has been commercialized for wastewater treatment, desalination, food, and beverages industries. The membrane has almost half replacement rate than that of conventional membranes and almost 10 times antibiofouling properties [23]. Table 3.1presents the major research carried out using 3D printing for water treatment.

3.7 Advanced applications of 3D printing for wastewater treatment

3.7.1 Ion-exchange membranes

The ion-exchange membranes are used in electrodialysis, biomedical, and microfluidic applications. Such membranes separate dissolved ions from salty water in the desalination process. Patterned membranes enhance ion transportation and diminish fouling. In a research work, an anion exchange membrane was fabricated using photolithographic approach and patterned membranes showed lower resistance, which is desirable. The layer wise manufacturing technique of 3D printing can be used for the fabrication of microbial desalination cell membranes and anodes for desalination. 3D printing may thus be useful for the manufacture of all components in membrane desalination to generate more energy in comparison to conventional materials.

3.7.2 Composite membranes

Polyamide composites are used for making composite membranes. The active film of the polyamide layer can be manufactured using 3D printing.

TABLE 3.1 Research work indicating the use of 3D printing for wastewater treatment.

3D printed part	3D printing technique	Material	Application	References
Spacers for membrane separation	Selective laser sintering	Polyamide 12 for triply periodic minimal surfaces spacers	Reverse osmosis, ultrafiltration	[20]
Spacers for membrane separation	Selective laser sintering	Polypropylene netted structures	Filtration using spiral wound membrane	[24]
Spacers for membrane separation	Fused deposition modeling, selective laser sintering, polyjet	Polyamide 12 white powder for spiral wound membrane modules	Filtration	[25]
Spacers for membrane separation	Polyjet	Urethane acrylate polymer	Nano filtration, reverse osmosis	[26]
Vibrating spacers for membrane separation	Selective laser sintering	Solid polyamide	Microfiltration	[27]
Spacer for membrane separation	Fused deposition modeling	Acrylonitrile butadiene styrene (ABS)	Reverse osmosis, ultrafiltration	[28]
Spacer for membrane separation	Digital light processing	Acrylate monomer column type design	Ultrafiltration	[29]
Thin film composite membranes	Inkjet printing and polymerization	Fluorinated diamine combined with m-phenylene diamine-based polyamide	Reverse osmosis and nanofiltration	[30]
Composite membrane	Multi-jet printing	ABS like substrate	Ultrafiltration	[31]

Continued

TABLE 3.1 Research work indicating the use of 3D printing for wastewater treatment.—cont'd

3D printed part	3D printing technique	Material	Application	References
Biocatalytic membrane	Ink jet printing	Polyvinyl alcohol and yeast cells	Advanced water treatment	[32]
Ferrate capsules	Lab made printer	Polyvinyl alcohol	Removal of pharmaceutical wastes	[33]
Fullerene inspired bio-carrier media	Selective laser sintering	Nylon	Biofilm reactor	[34]
Bio-filter media carrier	Polyjet	Liquid acrylate monomer resin	Moving bed biofilm reactor	[35]
Porous membrane	Digital light processing	Polydimethylsiloxane	Oil water separation	[36]

The layers of very small thickness can be produced and greater effectiveness is achieved for water flux and permeability. These spongy polymers possess good strength, high density, and are rigid resulting in greater salt rejection capacity during the reverse osmosis process. These fabrication processes are also environmental friendly and generate less waste.

3.7.3 Photocatalysis material

In heterogeneous photocatalysis, a semiconductor material gets activated when exposed to a light source at specific wavelength. The polluting get oxidized when they react with the activated semiconductor material. Photocatalytic materials like titanium dioxide increase the rate of wastewater treatment. The material must be immobilized on a substrate to avoid any secondary pollution from the catalysts and make reuse of photocatalysts. The photocatalyst substrates require more surface area to make the catalytic material adhere to the substrate and can be easily achieved if the substrate materials are 3D printed. The polymers used in 3D printing have low surface areas, vary in thermal stability and possess poor surface properties making them unsuitable for photocatalyis. 3D-printed polymers are thus used to make substrates.

3.7.4 Capsule/bio-carriers

Wastewater streams require post treatment to remove the effluents remaining after the treatment due to property and size constraints. Chemical oxidation using ferrates is done to remove these compounds. Production of ferrates is expensive, and they are unstable in humid environment. Capsulation makes ferrates stable, cost-effective, and efficient to remove contaminants. 3D is used to prepare polyvinyl alcohol capsules to encapsulate ferrates, which maintain efficiency for a long term. The limitation is the loading of ferrate to the capsule, which may be improved by making capsules with ferrates within them or making them as tablets with open-close structure prepared by multimaterial printer. In a recent study, 3D printing is used to design a novel fullerene design type bio-carrier made up of nylon, which has greater surface roughness and thus demonstrates good adhesion ability.

3.7.5 Moving bed biofilm reactors

MBBRs are used in wastewater treatment due to their simplicity of operation and efficiency. The performance depends on the formation of stable bio-films on the surface, which is affected by the design of the reactor and the operating conditions. 3D printing can produce complicated and good designs and ensure adhering of the bacteria to the surface of the reactor bed.

3.7.6 Oil–water separation substrate

Environmental pollution caused due to oil spillage during extraction, transportation, or exploitation requires porous membranes to clean up the water surfaces as they are efficient and can be recycled. The conventional membranes allow oil to pass through but prevent water at its surface. 3D-printed technology fabricates membranes with desirable properties and the researchers are focusing much in this area.

3.7.7 Solar absorbers

Wastewater can be effectively cleaned by the use of solar-driven water evaporation where solar illumination is projected on photothermal materials. The major challenge remains to fabricate materials, which can conveniently and efficiently convert solar illumination to thermal energy. 3D printing has helped overcome this challenge and has been able to fabricate photo thermal materials with patterns and designed architecture exhibiting high efficiency.

3.7.8 Substrates for dye degradation

The process of organic dye removal requires flexible materials. Carbon-based materials are used as adsorbents for the removal, and they make the process complicated due to the lack of flexibility in them. 3D printing has been applied in various research works to provide flexible materials with open structures that can be used for dye absorption while making the materials recyclable [12].

3.7.9 Heavy metal adsorption

Removal of heavy metals like mercury, cadmium, lead, and copper is essential for the human health. Materials like carbon nanotubes, activated carbon, and bio-inspired material are used for heavy metal removal through a porous media due to their low cost, high effectiveness, and biodegradability. 3D printing has the potential to fabricate such porous materials and overcome this challenge.

3.7.10 Micro devices

Microbial fuel cells oxidize organic pollutants in wastewater while producing power. Anodes of such system carry the bacteria and improvement in the anodes may significantly enhance the performance of the process. The porous carbon anodes have been prepared by 3D printing and carbonization [9].

3.8 Optimization and sustainability of 3D printing technology in wastewater treatment

3.8.1 Challenges for manufacturing of membranes

The 3D printing technology bears some challenges for the manufacturing of membranes, which include the following:

(i) Membrane configuration, which includes flat sheets, tubular, or hollow structures due to manufacturing limitations may be improved by using 3D printing techniques and optimized for obtaining better results.

(ii) The surface shapes and complex pore structures of membranes improve the micro-mixing and flow through the membranes. The membrane and energy requirement may be minimized by proper optimization techniques.

(iii) Membrane thickness can be controlled by 3D printing techniques and may be optimized by computational methods for improving process performance.

(iv) The pore size, distribution, and structure can be optimized in the CAD model to optimize the flux and separation efficiency.

(v) 3D printing technique requires more fabrication times and material requirement is also selective, which makes it costlier in comparison to other techniques. The reduction in the cost of printing materials is required, and optimization of the energy, printing material, and process requirement will help achieve the objective.

(vi) 3D printing techniques use less energy and solvents making the process environment friendly. The requirement of energy can be balanced with proper optimization and analysis.

3.9 Computational intelligent techniques for control and cost optimization of wastewater treatment

3.9.1 Artificial intelligence

Artificial intelligence is a powerful technique used to optimize wastewater and reduce the complexities and associated costs in the treatment process. The technology can analyze the historical and present treatment sensor and real-time data patterns to control the process, support decisions, and recommend actions for performance enhancement with increased efficiency. It can also improve the reliability and quality of the water and retain it for a greater time period. The AI models used are classified into single and combined methods. They can further be classified into machine learning techniques, search algorithms, and other techniques. Past research work focuses on the application of machine learning methods like artificial neural network (ANN), deep learning (DL) methods, principal component analysis (PCA), support vector machine (SVM), random forest (RF), that are used to handle nonlinear classification and regression analysis. The methods are applied for simulation and prediction of the performance of wastewater treatment process by removing pollutants, cost and energy efficiency, wastewater management, and reuse. ANN and DL are used to model the biological neurons and can solve multivariate nonlinear problems with proper training algorithm and right amount of data. ANN learns from the training data, relates it to the data points, and uses it for simulation, prediction, and optimization of the results. The search algorithms use methods like genetic algorithm and genetic programming, which find application as single AI technologies. Other models that have been used are fuzzy logic, neural fuzzy, expert system, artificial neural network-genetic hybrid algorithm models, model tree, data mining, clustering algorithm, Bayesian network, and particle swarm optimization. Adaptive neuro fuzzy inference system (ANFIS), a hybrid of neuro and fuzzy method, has also found application in the field of wastewater treatment [37].

The models have been developed using data from the experiments to simulate, predict, confirm, and optimize the parameters during wastewater treatment. The experimental data set is divided into three types or two types viz. trained, validated and tested, or trained and tested. The training data set is used for developing the model, validated is used to optimize the model, and testing data set is used to test the results in the prediction stage. For the management and control of wastewater, due to lack of reliable automation techniques, knowledge is collected from data and operator's experience for allowing the operators to understand and improve the performance of the process. Decision support systems and hybrid models have been used to implement adaptive software sensors. Many hybrid models have been proposed by the researchers for operating complex wastewater treatment processes, improving the process efficiency and reducing the energy consumed. Multiobjective optimization techniques have also been used to solve multiple conflicts like cost, quality, stability, etc. and have resulted in better performance than conventional methods. Sustainable development of wastewater treatment has gained interest, and AI has the desired potential to recover clean water, energy, and other nutrients during treatment. Rainfall index is also being used as an input and decisions depend on weather conditions.

3.9.2 Commonly used artificial intelligence methods

3.9.2.1 Machine learning methods

Machine learning (ML) solves four types of problems viz. classification, regression, dimensionality reduction, and clustering. The methods are suitable for solving classification and regression problems and build models, which are based on data patterns of the process. The method estimates the relationship between the input parameters and target output in a given data set of a system during the training stage. The new output is predicted with the use of the estimated nonlinear relationship of the parameters. ANN, DL, SVM, and RF handle nonlinear classification and regression analysis. Support vector machine (SVM) is a supervised learning approach to regression and classification. It uses structural risk minimization (SRM) principle and can be applied to wastewater treatment and other applications. Different kernels are used such as linear, RBF, Gaussian, and polynomial. The problem space and data set define the kernel types and function parameters. The problem inputs and calculated outputs establish the structure of the data set. The structure of the data set helps select the type of kernel suitable for the designed SVR model. ANN finds application in many fields of research due to its ability to train models that produce acceptable results in any problem space. The process causes delays as the training, and error reduction process is carried out with iterations. To avoid such delays extreme learning machines (ELM) are used which have single hidden layer feed forward networks. Such techniques have short learning times and thus reduce the process delays. To determine the input weights, the ELMs need more hidden neurons. The weight ELM (WELM) is more accurate than ELM. Sometimes, evaluation metrics methods are used to analyze the performance of the results in ML techniques. PCA solves dimensional reduction problems by processing high dimensional data. ANN and DL replicate the behavioral characteristics of biological neurons to process distributed parallel information with DL having a more complicated structure but providing greater prediction accuracy. SVM can be transformed into a convex optimization problem but require longer training time. RF is used for classification and regression analysis [23,38]. These models can predict complex systems with unknown interactions and are used in wastewater treatment for removing contaminants and membrane fouling. Machine learning has gained much influence on the aspects of 3D printing in design, fabrication, qualification to logistics, optimization, and enhancing the part quality. Table 3.2 shows the ML techniques applied for 3D printing process parameter optimization.

TABLE 3.2 Effective machine learning techniques for optimization of 3D printing process parameters.

3D printing process	Feature for optimization	Parameters for optimization	ML technique	References
Selective laser sintering	Density	Layer thickness, scan speed, spacing, laser power	ANN	[39]
Selective laser sintering	Build time	Volume, height, bounding box	ANN	[40]
Selective laser sintering	Shrinkage ratio	Layer thickness, scan speed, spacing, laser power, mode, temperature, time	ANN,GA	[41]
Selective laser sintering	Dimension	Layer thickness, scan speed, spacing, laser power	Radial basic function, C means, k means, ANN, pseudo inverse	[42]
Selective laser sintering	Tensile strength	Layer thickness, scan speed, spacing, laser power, powder temperature	ANN	[43]
Binder jet	Shrinkage rate, surface roughness	Layer thickness, drying time, heating power, printing saturation	ANN	[44]
Binder jet	Compressive strength, open porosity	Orientation, layer thickness, delay	Aggregated ANN	[45]

3.9.3 Application of intelligent technologies for wastewater treatment

3.9.3.1 *Conventional modeling versus intelligent models*

The content of pollutants, composition, shape, size, surface reflectivity, and variability in the nature of pollutants makes it difficult to predict models for wastewater treatment technology using conventional methods. The performance of membrane filtration is dependent upon the feed water quality and operating parameters. The conventional methods for modeling use numerical simulation for prediction. The models simulate the dynamic evolution of the particle size but may not consider the collision mechanism for all shape of pollutants making the predicted models inaccurate. The models are limited to analyze and predict membrane pollution under simple condition. The AI models are accurate for such conditions as they are not based on any assumptions under ideal conditions, are faster, convenient, and powerful, provide efficient method for membrane fabrication cost optimization. Machine learning is categorized into four groups of: supervised learning, unsupervised learning, semi-supervised learning, and reinforced learning. Supervised learning is used where the class label or the target predicted value is available in the training data. This permits the statistical model to predict class for new observations. In unsupervised learning, the class label or target predicted values are not known and the objective is to identify some underlying structure in the dataset using clustering or dimensional reduction. Semisupervised learning combines supervised and unsupervised learning algorithms while dealing with large amount of data, which is otherwise impractical and expensive. These models use labeled and unlabeled data and are simpler to use, cost-efficient, and show better performance. Training data used for reinforced learning algorithms indicate whether they are correct or not and they learn the good in interaction with the environment. Reinforced learning algorithms use exploration and exploitation. Exploration is taking action that has not been taken earlier and exploitation is generating the maximum reward. By using these two techniques, optimal solution may be obtained. The major challenge for data based modeling in complex industrial processes is the nature of the data, which are high dimensional, not linearly deposited, not stationary, having nonlinear relationships with faulty values, measurement rates, delays, and availability of sufficient data containing sufficient variation. For variable prediction, the methods are combined and used to gain the benefit offered by each. Iterative predictor weighting-partial least square (IPW-PLS) uses weighted prediction by collection of some regression models to predict the variable water quality parameters [46]. Smart technologies have been effectively applied to various drinking water treatment processes to forecast the coagulant dosage, predict the inequity in the potential of formation of disinfection by product, to forecast the control of fouling in the membrane, the technology may also be used in the preparation, and optimization of membranes and the prediction of water quality. AI tools offer a novel technique to further understand the multifaceted phenomena involved in process of water treatment. The intelligent techniques have the potential to help in the process of decision-making for drinking water treatment industry. On the other hand, the use of smart artificial intelligence technologies in treatment of drinking water process is not been used to a large extent as these technologies are still facing difficulty when applied to real life. The existing interpretable analysis model only focuses on local study, and additional investigation is required to explore coupling of neural network structure with treatment of water process. To develop a dynamic process, a combined model of ANN optimization and neural network is required. The other difficulty that the AI model encounters is that the supervised learning model is mainly used in process drinking water treatment and those models generally require prior knowledge. The precision in the prediction of the trained model based on past data may decrease when there is a significant change in the working conditions like water quality, operating parameters, etc. of the water treatment plant. In varying conditions, it is very difficult to predict the extent of reduction that is within an acceptable range.

The modern technologies applied to the autopilot solution with advanced analytics and predictive control for the biological processes of water reclamation plant is the recent development in the machine learning models. The main aim of autopilot solution is to get stable operating condition and maintain proper feed water quality in the reverse osmosis (RO) process. The second objective is to find an optimized nutrient removal, which reduces the aeration energy by calculating the exact influent load. The process operators and instrument anomalies are achieved with the solution as an early warning to any fluctuation in the parameters during the process. In order to achieve these goals, application of self-learning predictive control is preferred rather than reactive control. A conventional control method, which only uses proportional-integral derivative (PID) loops, can develop instabilities. PID loops are not effective for the long-term predictions of process parameters of water treatment plant. In wastewater treatment, the main aberrations are mostly caused by the daily pattern of the influent load. Predictive control method utilizes stable processes, which can accurately forecast the called output and cleverly adjusts for objective variations. Additionally, predictive analytics method also provides an opportunity for early detection of anomalies much before the sharp increase in the difference between the measured and predicted values. Aeration flow and ammonium concentration in the bioreactor can be predicted precisely in different advanced model developed in the recent time. Due to the high rate of aeration, there may be an abnormal increase in the ammonium concentration, which increases the concentrations of nitrate. To maintain the quality of drinking water, membrane filtration is used. On the other hand, fouling in the membrane is caused due to multiple variables, which are having nonlinear characteristics. Membrane filtration performance mainly depends on the quality of feed water and operating parameters. Disinfection of drinking water plays an important role in ensuring availability of useable water for healthy life, but some of the disinfectants tend to react with the elements of natural water and produce highly toxic compounds. Thus, the prediction of production potential of DBPs is essential. Fig. 3.1 shows the different AI and ML techniques used for 3D printing.

3.9.3.2 Fault detection and diagnosis

Fault detection in complex systems identifies the root cause of the abnormal conditions. When performed on historical data, it is known as statistical process monitoring. The technique identifies the points where the conditions change from normal to faulty basing on a statistical model considering normal operating conditions. The values which are close to the fault are checked for symptoms. The detection and diagnosis helps in root cause analysis. Fault detection techniques are being used for wastewater treatment considering the process data. Methods like PCA, kernel PCA which uses kernel functions, platinum concentration, which uses extreme learning machines based upon feed forward neural networks have been used. The potential causes for the faults may be determined before their occurrence by analyzing the patterns in the wastewater treatment process.

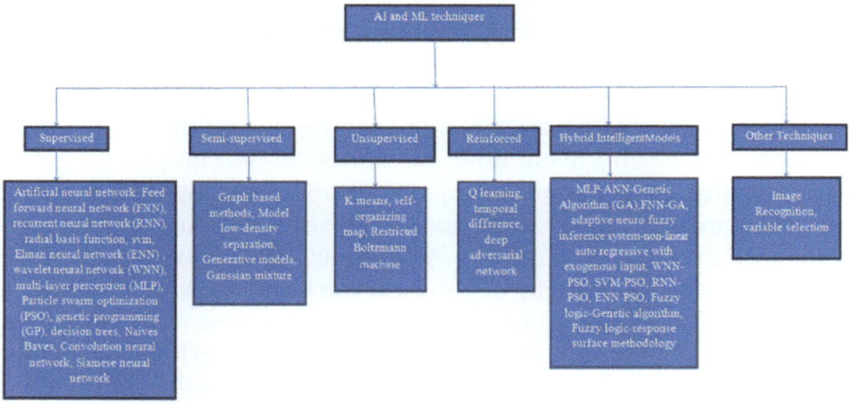

FIGURE 3.1 Artificial intelligence and machine learning models finding application in wastewater treatment.

3.9.3.3 *Part quality and process optimization for 3D printing in wastewater treatment using ML*

The relationship between process-structure-properties (PSP) in 3D printing is complex due to the difficulty in establishing the governing equations. Researchers are using ML algorithms to overcome the difficulties. Attention has been paid to the design of a surrogate system to match the dynamic characteristics of a model whose physical characteristics are not available [47]. In-situ monitoring can improve the reliability and repeatability of the 3D printed parts by closed loop feedback control using sensors. The anomaly detection can be performed using machine learning by a large data set of good and bad printing samples, which are used as new training data. The ML algorithms can predict no anomaly with 100% accuracy.

3.10 Future prospects and next generation applications of machine learning for wastewater treatment

(i) Combination of 3D printing and machine learning for wastewater treatment

The use of an integrated multi-process system is a potential future direction. As can be observed from the past research work, ML has been applied to enhance the performance of wastewater treatment technology. The efficiency of 3D printing has also increased by the use of ML techniques. But the application of ML methods for the optimization of different parameters during 3D printing of wastewater treatment components would be an attractive direction for study. The membrane fabrication technology needs improvement as printing the entire membrane with the required resolution using 3D printing technique is not yet possible. A hybrid method of 3D printing and electro-spraying for design and fabrication of membranes with high porosity will be more efficient if ML techniques are applied. The validation of various membrane designs and the simulation of parameters and analyses using computation fluid dynamics can be effectively done by the use of ML methods. Hybrid manufacturing of membranes using 3D printing and injection molding can result in faster production rate with sufficient cost-effectiveness while maintaining the flexibility in designs. The organic membrane materials, which are made up of polymers can be made using 3D printing technique and the ML, can be used to train the process of manufacture. By using various ML models, the optimum sizes of mesh to be used for membranes may be obtained so that the delay in getting the perfect sized membranes is eliminated. Also the errors can be minimized during the manufacture.

(ii) ML for cloud 3D printing

The cloud platform is advancing the processes to a greater extent. The server-based computational model includes hardware and software resources, which forms a comprehensive pool of resources and helps in sharing it to a public repository. ML can effectively evaluate and assess the 3D printers, membranes, and the resources in terms of accuracy, quality, cost, and time. It also helps in customizing the products and manages the resources using optimization algorithms and data acquisition methods.

(iii) Attack detection and security for data-driven models

The file sharing and cloud computing in 3D printing requires much of cyber security, which is a growing concern in the systems and wastewater treatment industries. ML can handle such situations and prevent malicious attacks. Supervised and unsupervised ML algorithms like k nearest neighbors (kNN), random forest, and anomaly detection may detect cyber-attacks of design, file transfers, and printing history with higher accuracy. Model detection algorithm prevents printing and sharing of restricted items.

3.11 Summary

The multitude of design and fabrication potential of 3D printing technology used for wastewater treatment has been discussed in this review article. The 3D-printed membranes used for wastewater treatment show better performance in comparison to conventional methods. The research trends in wastewater treatment have been discussed. The challenges faced during water purification processes have been discussed. The promising potential of artificial intelligence and machine learning techniques used for prediction and optimization during wastewater treatment has been discussed. Since there is greater variability of parameters in wastewater treatment process, to increase the efficiency, quality, and performance of the process, ML algorithms with better sensors, higher data acquisition rates, and high resolution must be used. The scope of using data driven methods for optimization of 3D-printed components for wastewater treatment have been discussed. The scope for future research has also been suggested which focuses on integrated learning models to optimize the process parameters during 3D printing component fabrication, and improve the sustainability of the design. The greater opportunities of using AI and ML methods for 3D printing of components used for wastewater treatment have been highlighted.

References

[1] T.D. Ngo, A. Kashani, G. Imbalzano, K.T.Q. Nguyen, D. Hui, Additive manufacturing (3D printing): a review of materials, methods, applications and challenges, Compos. B Eng. 143 (2018) 172–196, https://doi.org/10.1016/j.compositesb.2018.02.012.

[2] M.N. Nadagouda, M. Ginn, V. Rastogi, A review of 3D printing techniques for environmental applications, Curr. Opin. Chem. Eng. 28 (2020) 173–178, https://doi.org/10.1016/j.coche.2020.08.002.

[3] H.A. Balogun, R. Sulaiman, S.S. Marzouk, A. Giwa, S.W. Hasan, 3D printing and surface imprinting technologies for water treatment: a review, J. Water Proc. Eng. 31 (2019), https://doi.org/10.1016/j.jwpe.2019.100786.

[4] E.M. Dogo, N.I. Nwulu, B. Twala, C. Aigbavboa, A survey of machine learning methods applied to anomaly detection on drinking-water quality data, Urban Water J. 16 (3) (2019) 235–248, https://doi.org/10.1080/1573062X.2019.1637002.

[5] L. Zhao, T. Dai, Z. Qiao, P. Sun, J. Hao, Y. Yang, Application of artificial intelligence to wastewater treatment: a bibliometric analysis and systematic review of technology, economy, management, and wastewater reuse, Process Saf. Environ. Protect. 133 (2020) 169–182, https://doi.org/10.1016/j.psep.2019.11.014.

[6] M. Castrillo, Á.L. García, Estimation of high frequency nutrient concentrations from water quality surrogates using machine learning methods, Water Res. 172 (2020), https://doi.org/10.1016/j.watres.2020.115490.

[7] G. Crini, E. Lichtfouse, Advantages and disadvantages of techniques used for wastewater treatment, Environ. Chem. Lett. 17 (1) (2019) 145–155, https://doi.org/10.1007/s10311-018-0785-9.

[8] C. Fonseca Couto, L.C. Lange, M.C. Santos Amaral, A critical review on membrane separation processes applied to remove pharmaceutically active compounds from water and wastewater, J. Water Proc. Eng. 26 (2018) 156–175, https://doi.org/10.1016/j.jwpe.2018.10.010.

[9] A.N. Ahmed, F.B. Othman, H.A. Afan, R.K. Ibrahim, C.M. Fai, S. Md, M. Hossain, A. Ehteram, Elshafie, Machine learning methods for better water quality prediction, J. Hydrol. 578 (2019) 124084.

[10] H. Guo, K. Jeong, J. Lim, J. Jo, Y.M. Kim, J.p. Park, J.H. Kim, K.H. Cho, Prediction of effluent concentration in a wastewater treatment plant using machine learning models, J. Environ. Sci. (China) 32 (2015) 90–101, https://doi.org/10.1016/j.jes.2015.01.007.

[11] M. Padaki, R. Surya Murali, M.S. Abdullah, N. Misdan, A. Moslehyani, M.A. Kassim, N. Hilal, A.F. Ismail, Membrane technology enhancement in oil-water separation. A review, Desalination 357 (2015) 197–207, https://doi.org/10.1016/j.desal.2014.11.023.

[12] L.D. Tijing, J.R.C. Dizon, I. Ibrahim, A.R.N. Nisay, H.K. Shon, R. C, Advincula, 3D printing for membrane separation, desalination and water treatment, Appl. Mater. Today 18 (2020).

[13] V. Yangali-Quintanilla, S.K. Maeng, T. Fujioka, M. Kennedy, G. Amy, Proposing nanofiltration as acceptable barrier for organic contaminants in water reuse, J. Membr. Sci. 362 (1–2) (2010) 334–345, https://doi.org/10.1016/j.memsci.2010.06.058.

[14] B. Van der Bruggen, M. Mänttäri, M. Nyström, Drawbacks of applying nanofiltration and how to avoid them: a review, Separ. Purif. Technol. 63 (2) (2008) 251–263, https://doi.org/10.1016/j.seppur.2008.05.010.

[15] E. Curcio, E. Drioli, Membrane distillation and related operations - a review, Separ. Purif. Rev. 34 (1) (2005) 35–86, https://doi.org/10.1081/SPM-200054951.

[16] A. Alkhudhiri, N. Darwish, N. Hilal, Treatment of saline solutions using air gap membrane distillation: experimental study, Desalination 323 (2013) 2–7, https://doi.org/10.1016/j.desal.2012.09.010.

[17] T.Y. Cath, V.D. Adams, A.E. Childress, Experimental study of desalination using direct contact membrane distillation: a new approach to flux enhancement, J. Membr. Sci. 228 (1) (2004) 5–16, https://doi.org/10.1016/j.memsci.2003.09.006.

[18] R.W. Baker, Membrane Technology and Applications, Wiley Online Library, Hoboken NJ USA, 2012.

[19] H. Dommati, S.S. Ray, J.C. Wang, S.S. Chen, A comprehensive review of recent developments in 3D printing technique for ceramic membrane fabrication for water purification, RSC Adv. 9 (29) (2019) 16869–16883, https://doi.org/10.1039/c9ra00872a.

[20] N. Sreedhar, N. Thomas, O. Al-Ketan, R. Rowshan, H. Hernandez, R.K. Abu Al-Rub, H.A. Arafat, 3D printed feed spacers based on triply periodic minimal surfaces for flux enhancement and biofouling mitigation in RO and UF, Desalination 425 (2018) 12–21, https://doi.org/10.1016/j.desal.2017.10.010.

[21] T. Luukkonen, J. Yliniemi, H. Sreenivasan, K. Ohenoja, M. Finnilä, G. Franchin, P. Colombo, Ag- or Cu-modified geopolymer filters for water treatment manufactured by 3D printing, direct foaming, or granulation, Sci. Rep. 10 (1) (2020), https://doi.org/10.1038/s41598-020-64228-5.

[22] P. He, X. Tang, L. Chen, P. Xie, L. He, H. Zhou, D. Zhang, T. Fan, Patterned carbon nitride-based hybrid aerogel membranes via 3D printing for broadband solar wastewater remediation, Adv. Funct. Mater. (2018) 1801121, https://doi.org/10.1002/adfm.201801121.

[23] 3D printed membrane could revolutionise water filtration, Pump. Ind. Anal. (2014) 2–3, https://doi.org/10.1016/s1359-6128(14)70344-1.

[24] W.S. Tan, C.K. Chua, T.H. Chong, A.G. Fane, A. Jia, 3D printing by selective laser sintering of polypropylene feed channel spacers for spiral wound membrane modules for the water industry, Virtual Phys. Prototyp. 11 (3) (2016) 151–158.

[25] W.S. Tan, S.R. Suwarno, J. An, C.K. Chua, A.G. Fane, T.H. Chong, Comparison of solid, liquid and powder forms of 3D printing techniques in membrane spacer fabrication, J. Membr. Sci. 537 (2017) 283–296.

[26] A. Siddiqui, N. Farhat, S.S. Bucs, R.V. Linares, C. Picioreanu, J.C. Kruithof, M.C.M. Van Loosdrecht, J. Kidwell, J.S. Vrouwenvelder, Development and characterization of 3D-printed feed spacers for spiral wound membrane systems, Water Res 91 (2016) 55–67.

[27] Y.Z. Tan, Z. Mao, Y. Zhang, W.S. Tan, T.H. Chong, B. Wu, J.W. Chew, Enhancing fouling mitigation of submerged flat-sheet membranes by vibrating 3D-spacers, Sep. Purif. Technol. 215 (2019) 70–80.

[28] A. Shrivastava, S. Kumar, E.L. Cussler, Predicting the effect of membrane spacers on mass transfer, J. Membr. Sci. 323 (2) (2008) 247–256.

[29] S.M. Ali, A. Qamar, S. Kerdi, S. Phuntsho, J.S. Vrouwenvelder, N. Ghaffour, H.K. Shon, Energy efficient 3D printed column type feed spacer for membrane filtration, Water Res 164 (2019) 114961.

[30] S. Badalov, Y. Oren, C.J. Arnusch, Ink-jet printing assisted fabrication of patterned thin film composite membranes, J. Membr. Sci. 493 (2015) 508–514.

[31] S. Mazinani, A. Al-Shimmery, Y.M.J. Chew, D. Mattia, 3D printed fouling-resistant composite membranes, ACS Appl. Mater. Interfaces (2019).

[32] Y. Chen, P. Gao, M.J. Summe, W.A. Phillip, N. Wei, Biocatalytic membranes prepared by inkjet printing functionalized yeast cells onto microfiltration substrates, J. Membr. Sci. 550 (2018) 91–100.

[33] M. Czolderova, M. Behul, J. Filip, P. Zajicek, R. Grabic, A.V. Stanova, M. Gál, K. Kerekes, J. Hives, J. Ryba, M. Rybanska, P. Brandeburova, T. Mackulak, 3D printed polyvinyl alcohol ferrate (VI) capsules: Effective means for the removal of pharmaceuticals and illicit drugs from wastewater, Chem. Eng. J. 349 (2018) 269–275.

[34] Y. Dong, S.Q. Fan, Y. Shen, J.-X. Yang, P. Yan, Y.P. Chen, J. Li, J.S. Guo, X.M. Duan, F. Fang, S.Y. Liu, A novel bio-carrier fabricated using 3D printing technique for wastewater treatment, Sci. Rep. 5 (2015) 12400.
[35] O. Elliott, S. Gray, M. McClay, B. Nassief, A. Nunnelley, E. Vogt, J. Ekong, K. Kardel, A. Khoshkhoo, G. Proano, D.M. Blersch, A.L. Carrano, Design and manufacturing of high surface area 3D-printed media for moving bed bioreactors for wastewater treatment, J. Contemp. Water Res. Educ. 160 (1) (2017) 144–156.
[36] J. Lv, Z. Gong, Z. He, J. Yang, Y. Chen, C. Tang, Y. Liu, M. Fan, W.M. Lau, 3D printing of a mechanically durable super hydrophobic porous membrane for oil–water separation, J. Mater. Chem. 5 (24) (2017) 12435–12444.
[37] A. Bernardelli, S. Marsili-Libelli, A. Manzini, S. Stancari, G. Tardini, D. Montanari, G. Anceschi, P. Gelli, VenierWater Science & Technology Real-Time Model Predictive Control of a Wastewater Treatment Plant Based on Machine Learning, 2020.
[38] L. Li, S. Rong, R. Wang, S. Yu, Recent advances in artificial intelligence and machine learning for nonlinear relationship analysis and process control in drinking water treatment: a review, Chem. Eng. J. 405 (2021) 126673, https://doi.org/10.1016/j.cej.2020.126673.
[39] X. Shen, J. Yao, Y. Wang, J. Yang, Density prediction of selective laser sintering parts based on artificial neural network, in: International symposium on neural networks, Springer, Berlin, 2004.
[40] J. Munguia, J. Ciurana, C. Riba, Neural-network-based model for build-time estimation in selective laser sintering, Proc. Inst. Mech. Eng. B: J. Eng. 223 (8) (2009) 995–1003.
[41] W. Rong-Ji, L. Xin-hua, W. Qing-ding, W. Lingling, Optimizing process parameters for selective laser sintering based on neural network and genetic algorithm, J. Adv. Manuf. Technol. 42 (11–12) (2008) 1035–1042.
[42] X. Li, J. Dong, Y. Zhang, Modeling and Applying of RBF Neural Network Based on Fuzzy Clustering and Pseudo-Inverse Method, in: International Conference on Information Engineering and Computer Science, 2009, pp. 1–4.
[43] C. Wang, N. Jiang, Z. Chen, Y. Chen, Q. Dong, Prediction of sintering strength for selective laser sintering of polystyrene using artificial neural network, J. Donghua University 32 (2015) 825–830.
[44] H. Chen, Y.F. Zhao, Learning algorithm based modeling and process parameters recommendation system for binder jetting additive manufacturing process, in: ASME international design engineering technical conferences and computers and information in engineering conference, 2015.
[45] M. Asadi-Eydivand, M. Solati-Hashjin, A. Fathi, M. Padashi, N. Azuan Abu Osman, Optimal design of a 3D-printed scaffold using intelligent evolutionary algorithms, Appl. Soft Comput. 39 (2016) 36–47.
[46] G.D. Goh, S.L. Sing, W.Y. Yeong, A review on machine learning in 3D printing: applications, potential, and challenges, Artif. Intell. Rev. 54 (1) (2021) 63–94, https://doi.org/10.1007/s10462-020-09876-9.
[47] A. Yusuf, A. Sodiq, A. Giwa, J. Eke, O. Pikuda, G. De Luca, J.L. Di Salvo, S. Chakraborty, A review of emerging trends in membrane science and technology for sustainable water treatment, J. Clean. Prod. 266 (2020) 121867, https://doi.org/10.1016/j.jclepro.2020.121867.

CHAPTER

4

An overview of water pollutants in present scenario

Santhosh Naraseeyappa Kuchangi, Manohara Halanur Mruthunjayappa, and Nataraj Sanna Kotrappanavar

Sustainable Materials and Processes Lab, Centre for Nano & Material Sciences, Jain Global Campus, Jain University, Bangalore, Karnataka, India

4.1 Introduction

Water comprises most of living tissues as a fundamental necessity of life. However, the global demand for clean water is increasing at a fast pace due to population growth, urbanization, agricultural expansion, industrialization, and climate change. Globally, freshwater occurs as surface water and groundwater naturally [1]. Though groundwater contributes only 0.6% of the total water resources on earth, it is the major and the preferred source of drinking water in rural as well as urban areas, particularly in the developing countries like India. It caters to 80% of the total drinking water requirement and 50% of the agricultural requirement in rural India. Fig. 4.1 explains about global water consumption in all provinces like agriculture, industries, and domestic. In this, 69% of the world's freshwater was used for agricultural purpose, with the remaining 23% and 8% going to the industrial and domestic applications, respectively [2].

Thus, water pollution is one of the greatest ecological issues making significant issues to the living creatures. The pollution of water is brought about by many inorganic and natural pollutants in water, some of which are profoundly tireless, harmful, and cancer-causing. More often, ground water is brackish, saline, or contaminated with excess geogenic contaminants, such as fluoride, arsenic, iron, microorganisms, etc. On the other side, many aquifers in various parts of the world have been found to be contaminated with organic, inorganic, and heavy metal ions due to

3D Printing Technology for Water Treatment Applications
https://doi.org/10.1016/B978-0-323-99861-1.00001-1

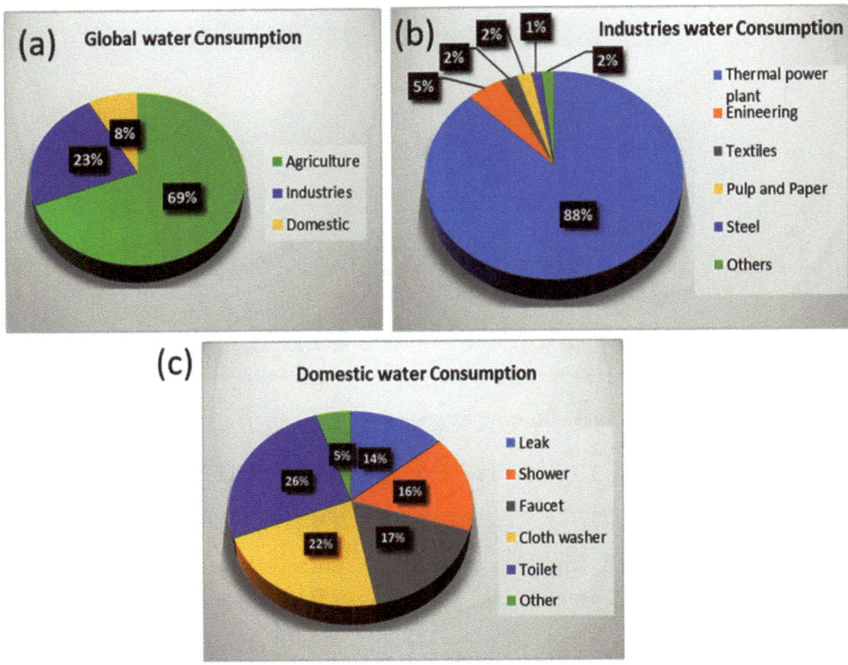

FIGURE 4.1 Detailed water consumption in all sectors like agriculture, industries, and domestic, resulting water consumptions are follows. (A) Global water consumption, (B) industries water consumption, and (C) domestic water consumption.

natural disasters and human activity [3]. Major harmful pollutants are pesticides, polynuclear fragrant hydrocarbons, plasticizers, phenols, polybrominated diphenyl ethers, polychlorinated biphenyls, and, drug deposits, while essential inorganic pollutants include arsenic, lead, cadmium, mercury, chromium, and numerous other (poisonous) metals, alongside supplements like fluoride, nitrate, and phosphate. Environmental pollutants, industry effluents, populace development, and also, long-term drought seasons cause a significant issue around the world [4]. New pollutants continue to arise and present extreme well-being and scientific challenges. It is resourced that around four billion people overall experience to have no or little admittance to spotless and cleaned water supply, and millions of people passed on serious waterborne sicknesses yearly. These factual figures are required to fill in the short future, as expanding water pollution due to overpowering release of micro-pollutants and contaminants substances into the characteristic water cycle. In view to restrict the deteriorating of clean water deficiency, improvement of progressed with minimal expense and high productivity water treatment advances to treat the wastewater is attractive [5].

The storage of rain water for every day exercises and expanding the catchment area limit with respect to storm water are simply a couple of models that could resolve the issues in present moment. Water industries in some dry zones with bountiful of daylight, less precipitation, and long-term droughts spell have a test to look for viable water resource. Thus, expulsion of different harmful substances from water and wastewater with recycling the water has been a center of interest for numerous researchers over the past many years. Various technologies are being used to remove these contaminants from water, but still the problem has not been rooted out [6]. As industrial and anthropogenic exercises are representing a phenomenal danger to the common habitat, the evacuation and immobilization of impurities are of essential significance for the security of environments. However, in order to separate pollutants from water, it is necessary to have knowledge of pollutants. Thus, the present book chapter fed light on the classification and nature of various pollutants present in aqueous system.

4.2 Source of pollutants

4.2.1 Point source of pollution

Contamination beginning from a solitary, recognizable source, for example, a release pipe from a plant or sewage plant, is called point-source contamination. Contamination that does not begin from a solitary source, or point, is called nonpoint-source contamination. Fluid, strong, and airborne releases from point sources just as toxins from nonpoint sources may go either into surface water or into the ground. The capacity for these poisons to arrive at surface water or groundwater is improved by the measure of water accessible from precipitation or irrigation. Point-source contaminations in surface water and groundwater are normally found in a tuft that has the most elevated concentration of the toxin accessible source (like the finish of a line or an underground infusion framework) and decreasing focuses farther away from the source. The different kinds of point-source toxins found in waters are just about as fluctuated as the sorts of business, industry, rural, and metropolitan sources that produce them. The crude materials and squanders may incorporate poisons like solvents, oil-based goods (like oil and fuel), or substantial metals [7].

Point sources of contamination from agribusiness may incorporate creature taking care of activities, creature squander treatment tidal ponds, or capacity, handling, blending, and cleaning zones for pesticides, manures, and oil. City point sources may incorporate wastewater treatment plants, landfills, utility stations, and engine pools. The absolute most steady point-source contaminations in groundwater are unpredictable natural mixtures, which incorporate fabricated and refined harmful substances like solvents, oils, paint, and fuel items. As a rule, it takes just a modest quantity of these synthetic substances to raise wellbeing concerns. For instance, roughly 4 L (around 1 gallon) of unadulterated trichloroethylene, a typical dissolvable, will debase more than one billion liters (300 million gallons) of water [7,8].

Point sources of contamination are brought about by releasing fluids directly into the water, for example, an organization that ousts defiled water or harmful solids blended directly with water into the ocean or stream. This makes the water virulent, from time to time coming about in demise, for fish and other oceanic animals. By and by, creatures additionally drink this water, which additionally bring to them weakness or demise [9]. It can like-wise influence to people. In created nations, individuals at this point do not rely upon drinking water from the stream or waterway. Circuitous water contamination is not brought about by the presentation of toxins straightforwardly into the water yet by those that end up there. A model incorporates compost furthermore, pesticide synthetic compounds that are washed gradually through the dirt and discover their direction into groundwater and afterward into different watercourses. Also, air contamination can cause corrosive downpour to tumble to the ground, which can be very unsafe to untamed life, counting dirtying lakes, streams, and shores, and make the water lethal for those animals that live in and close to it [10]. Regardless of whether it is immediate or backhanded water contamination, the outcomes can in any case be the same, for example, illness and potentially demise to any living thing that lives in it or takes in water. In this way, it is basic that people and organizations make moves to decrease their degrees of contamination and diminish their ecological effect. Organic compound comprises carbon, hydrogen, oxygen, nitrogen, and sulfur. Organic compound discharged from sewage, metropolitan waste water, modern wastewater, and also, agricultural waste. Evolving nations are analyzed about defilement of contamination chemicals such as nitrite, ammonium nitrate, and substantial metals in drinking water. The undeniable degrees of inorganic nitrogen pollutants (nitrate, nitrite, and ammonium) and inorganic phosphates in waterway water coming about because of drinking water from agricultural fields, releasing city/modern sewage, and so forth lead to numerous medical problems. Nitrite is cancer-causing in nature, which builds the danger of stomach, liver, and esophageal disease and can prompt undeniable degrees of ammonium in the body [11].

4.2.1.1 *Sewage wastewater*

Sewage wastewater is a kind of wastewater that is liberated by a local area of people. It is described by volume or pace of stream, state of being, substance and harmful constituents, and its bacteriologic status (which life forms it contains and in what amounts). It comprises generally of greywater (from sinks, baths, showers, dishwashers, and garments washers), blackwater (the water used to flush latrines, joined with the human waste that it flushes away), cleansers, and bathroom tissue (less so in districts where bidets are broadly utilized rather than paper) [12]. Legitimate assortment and safe, aggravation-free removal of the fluid wastes of a local area are lawfully perceived as a need in an urbanized, industrialized society. Sewage typically goes from a structure's pipes either into a sewer, which will convey it somewhere else or into an on the location sewage office (of which there are numerous sorts). Regardless of whether it is joined with surface encroachment in the sewer relies upon the sewer plan (sterile sewer or consolidated sewer). Actually, most wastewater created worldwide remaining parts untreated, causing inevitable water contamination, the 90% of all wastewater produced is released into the climate untreated. In many nonindustrial nations, the majority of home grown and modern wastewater is released with no treatment [13]. The necessities of researching conceivable imaginative courses are clear for sewage sludge valorization. The utilization of sewage sludge to eliminate toxins present in wastewater would offer both an administration choice for sewage sludge and furthermore give increase valorization as a minimal expense precursor for initiated carbon creation. As the costs caused by a sewage plant in the treatment of sludges address a significant level of complete expenses, the utilization of these sludge in the treatment of waters entering the actual plant would decrease costs [14].

4.2.1.2 Pharmaceutical industry

Pharmaceuticals are pollutants of developing analyze but have been found in trace quantities in water sources globally. Pharmaceuticals are among the extensively utilized chemical classes in nutrition, diagnostic aids, remedy, treatment, and preventive medication. Pharmaceutical companies, felonious medications, chemicals, hormones, and antibiotics are essential emerging toxins arising foreign substances because of their monstrous presence in aquatic environment, fresh water defilement, and conceivable adverse consequences on biodiversity and human being [15]. Sewage weakening after pharmaceuticals are released into water body adequately diminishes feasible environment danger. Along with perforation by personal use, pharmaceuticals were too utilized in livestock, domesticated animals, and fishery. Various drugs are broadly given to domesticated animals to limit disorders in such animals just as to improve the weight of the mammals [16]. Considering the reality that around 3000 chemical substances are being utilized in medicinal products, simply a little level of these chemical compounds has been tested in the field. Researchers and assessors have inspected regardless of whether presence of ng/L dose present in the climate might prompt antagonistic effects in biodiversity or people. The predominant ways for pharmaceuticals into the atmosphere were by means of individual discharge, the left-over of extra medications and cultivating application. A huge assortment of pharmaceuticals connected with wastewater releases were recognized in groundwater and surface water. Scientific examinations have shown that conventional treatment plants, essentially depending on utilizing miniature—living beings, are deficient to in a real sense execute the sorts of natural substances as a result of their muddled sub-atomic design and restricted qualities, and furthermore the level of expulsion was simply up to 10%. Pharmaceuticals that are more usually reported in wastewater and drinking water include: antibiotics, diclofenac, antacids, clofibric acid, steroids, antidepressants, ciprofloxacin, analgesics, salicylic acid, anti-inflammatory drugs, propranolol, antipyretics, beta blockers, fluoxetine, nitroglycerin, lipid-bringing down drugs, sedatives, and stimulants. The chemicals such as acetaminophen, ibuprofen, and enalapril were completely taken out generally during treatment techniques second class of reagents like lipid controllers were not entirely removed, despite the fact that their dosages were significantly more modest after treatment than those in powerful ones [16,17]. A group 3 of substances such as anxiolytics, antibiotics, and macrolide were not recognized in powerful profluent but rather were found in sewage water of effluents. Last class like simvastatin, sulfamethazine, and paroxetine, was not ever found in powerful gushing or wastewater emanating. Illicit medications have been at first distinguished in aquatic environment in 2004. Recently it is found in squander water, groundwater, and furthermore, surface water. Dosages in surface water are typically small, anyway a few wastewaters may include μg/L. Benzoylecgonine, cocaine, 2-ethylidene-1, methadone, and their metabolites, 5-dimethyl-3, 3-diphenylpyrrolidine were all the more normally saw in groundwater, anyway at restricted ng/L fixations. Triclosan antimicrobial is an essential segment close by hand soaps and toothpastes and is usually present in the atmosphere. Hormones were even fundamental toxins in the biological system as a result of their limit with respect to androgenic and estrogenic effect on biodiversity [18]. Organic or inorganic hormones incorporate 17α-estradiol, 17β-estradiol, estrone, equilenin, estriol, mestranol, equiline, and norethindrone, which can incorporate the atmosphere by means of cultivating and processed wastewater but still not diverged entirely. Photo-catalysis using the titanium dioxide semiconductor additionally demonstrated a fruitful clean water treatment strategy. The viability of this oxidative method additionally has been appeared for the debasement of pharmaceutical items. The arising pollutants in the pharmaceuticals are threat even at low focus to the wild-life, aquatic animals, human, and environment [19].

4.2.1.3 *Personal care products*

Personal care products are an assortment of conventional family chemical compounds that are used for wellbeing, beauty, and cleaning. Fancy cosmetics, skin products, hair care products, cleaning products, and fragrances covered personal care products. Skincare and personal care things were used in enormous amount of all throughout the world; as a result of persistently everyday use, these are released into the air in very enormous volumes continually [20]. Because of individual activities, the measure of emerging pollutants discharged into the air is rising each day and addresses the remarkable utilization of a wide assortment of things, like beauty and personal care things. The vast majority of these materials are bioactive and ordered by their ability for maintenance and bio-accumulation, making a danger to the environment and human public. Arising toxins are frequently used in skin health management products and can at last enter the environment at levels that could harm the people [21]. The paramount concentrations of emerging pollutants were found in beauty products, along with TiO_2 and ZnO nanoparticles fetching major out to be driving critical possible toxins. The most emerging pollutants in personal care products have been listed: polydimethylsiloxane, nano titanium dioxide, butylated hydroxylanisole, insect repellents, microplastics, butylated hydroxytoluene, disinfectants pollutants like triclosan and triclocarban, fragrances pollutants like galaxolide, tonalide, celestolide and phantolide, preservatives diethyl phthalate, nano zinc oxide, benzophenone, parabens, octinoxate methoxycinnamate, and butylparaben [22]. Personal care products, for example, shampoos, cleaning products, hair care products, washing products, and cosmetics procedures additionally contain polycyclic musks as modest fragrances. Its execution basically outwardly of the human skin improves its probability of higher event all through the environment without metabolic changes give the list of arising pollutants in personal care products and its properties since beautifiers are the one which we use in standard way, subsequently the follow measure of this toxins may make harm human, animals, and the environment [23].

4.2.1.4 *Pesticides*

Pesticides contain a vast diversification of discrete with explicit physical and chemical properties, which can be utilized in cultivating exercises to limit or restrain the spread of dangerous insects, weeds, and micro-organism like parasites and microbes. In extensive, pesticides are classified into four classes: insecticides, fungicides, herbicides, and bactericides. Its adaptation to the marine biological systems is accomplished through overflow from its site of utilization, and these contaminations may bio-aggregate in living things, vegetation, or sediments relying upon the substance solubility [24]. At the point when an emerging pollutant is brought into or set up in the agro-ecosystem, it will confront exactly the same destiny and transportation appliance components, which occur for some other pollutant variety. The emerging toxicity substances can be debased by organic, physical, or chemical strategies, they can bind to soil debris, they can be gotten by the roots, they can dissolve to groundwater, or they can be conveyed by precipitation and water system water to the surface waters. The degree to that every one of these mechanisms can happen relies upon the fundamental actual attributes of the arising poisons, like solubility; fascination of emerging pollutants toward organic matter and other substance soil; and reactivity of the substance, and furthermore the attributes of soils and environment [25]. Endosulfan, hexachlorocyclohexane, and dichlorodiphenyltrichloroethane (DDT) were among the often routinely operated pesticides, with the last two mixtures representing 67% of by and large pesticide utilization. Different items contain phorate, chlorpyriphos, methyl parathione, bentazone, cyanazine, simazine, atrazine, terbuthylazine, diazinon, alachlor, dimethoate, and phosphamidone. The arising pollutants in Pesticides and its wide properties. These are the distinctive arising contaminants risen up out of pesticides. Out of all pesticides pollutants almost 67% was covered by hexachlorocyclohexane and dichlorodiphenyltrichloroethane [26].

4.2.1.5 Oil pollution

The oil pollution is the arrival of a fluid petrol hydrocarbon into the climate, particularly the marine biological system, because of human action and is a type of contamination. The term is generally given to marine oil pollution, where oil is delivered into the sea or waterfront waters, however pollution may likewise happen ashore [27]. Oil pollution might be because of arrivals of unrefined petroleum from big haulers, seaward stages, penetrating apparatuses, and wells, just as refined oil-based commodities (like gas, diesel) and their side-effects, heavier powers utilized by huge ships, for example, fortification fuel, or the spill of any oily deny or waste oil. Oil pollution infiltrates into the construction of the plumage of birds and the hide of warm-blooded animals, lessening its protecting capacity and making them more vulnerable against temperature changes and substantially less light in the water. Clean-up and recuperation from an oil pollution are troublesome and relies on numerous variables, including the kind of oil polluted, the temperature of the water (influencing dissipation and biodegradation), and the sorts of shorelines and sea shores included [28].

The polluted oil can influence creatures and direct from the oil and from the reaction or cleanup process. There is no reasonable connection between the measure of oil in the oceanic climate and the presumable effect on biodiversity. A more modest pollution at some unacceptable time/wrong season and in a sensitive climate may demonstrate significantly more hurtful than a bigger pollution at some other point of the year in another or even the equivalent environment [29]. Oil enters into the design of the plumage of birds and the hide of vertebrates, decreasing their protecting capacity and making them more powerless against temperature changes and substantially less light in the water. Furthermore, oil pollution can likewise graze air quality. The synthetic substances in raw petroleum are generally hydrocarbons that contain poisonous synthetic compounds like benzenes, toluene, poly-fragrant hydrocarbon, and oxygenated polycyclic sweet-smelling hydrocarbons [30]. These synthetic substances can present antagonistic well-being impacts when being breathed in into human body. These synthetics can be oxidized by oxidants in the air to frame fine particulate matter after they vanish into the atmosphere. These particulates can enter lungs and convey harmful synthetics into the human body. Consuming surface oil can likewise be a hotspot for contamination, for example, residue particles. During the clean-up and recuperation measure, it will likewise produce air toxins like nitric oxides and ozone from ships. Secondary containment methods prevent releases of oil or hydrocarbons into the environment. Digging for oils scattered with cleansers and different oils denser than water [31].

4.2.1.6 *Radioactive pollution*

Conventionally radioactive material comes from earth crust and dissolves in surface drinking water. Anthropogenic radioactive material is transmitted from atomic force plant, atomic weapons testing, and production and utilization of radioactive material. For the most part, radionuclides showed up in drinking water have arrangement of uranium, thorium, and aluminum and with normally happening materials of radium, uranium, and the radioactive gas radon. These impurities cause perilous impact on human being. Radium causes bone malignant growth. Uranium likewise causes cancer in bone, toxic effect on kidney [32].

Atomic energy is premise of clean-air and carbon free electricity, but the radioactive particle tainting that is emerged from, e.g., tailings and stack spillage survives from uranium mining (like ^{226}Ra particles), nuclear fusion derivatives (like ^{90}Sr), atomic weapon tests, clinical exploration and spillage, and incidental blasts in the atomic reactors can make a drawn-out impact as a momentous danger to human being [33]. Thermal energy stations and modern effluents are the source of radioactive contamination in water. The polluting cycle released tremendous energy, and this energy goes as waves, and surprisingly smidgen of this is sufficient to cause harm in human cells, tissues, and organs. These destructive waves cannot be dealt with truly, synthetically, and naturally. Radioactive impacts persevere in the climate for broadened length anticipate on the natural half-life [34]. Topographically, long-term statement of all components including radioactive minerals gets presented to the topographical specialists subsequent to enduring and disintegration and radioactive components get assembled with time. The enduring rate relies upon the mineral synthesis, natural condition, and land specialists additionally. Least hint of radioactive component related with all groundwater relying on the geographical stone and layers of that space. Radionuclides either normal or fake have been available in water due to their sweep size and rot results [35].

4.2.2 Nonpoint source of pollution

Nonpoint-source contamination happens as water gets across the land or through the ground and gets common and human-made poisons, which would then be able to be saved in lakes, waterways, wetlands, waterfront waters, and even groundwater. The water that conveys nonpoint-source contamination may begin from regular cycles like rainfall or snowmelt or from human activities, for example, crop water system or grass support [36]. Nonpoint-source contamination is normally discovered fanned out all through a huge region. It is regularly hard to follow the specific beginning of these contaminations since they result from a wide assortment of human exercises on the land just as normal attributes of the dirt, environment, and topography. The most well-known nonpoint-source poisons are silt, supplements, microorganisms, and toxics. Sediment can corrupt water quality by polluting drinking water supplies or silting in bringing forth reason for fish and other oceanic species. Supplements, microorganisms, and other poisonous substances can be dangerous to human well-being and oceanic life [37]. Microscopic organisms, microorganisms, and supplements (nitrogen and phosphorus) are normal nonpoint-source contaminations from rural regions and private pet squanders. These toxins are likewise found in regions where there is a high thickness of septic frameworks or where the septic frameworks are defective or not looked after as expected. Different toxins from nonpoint sources incorporate salt from water system practices or street de-icing and corrosive seepage from abandoned mines [38].

4.2.2.1 Agriculture

Agricultural contamination alludes to biotic and abiotic side-effects of cultivating enumerate that bring about pollution or debasement of the climate and encompassing biological systems and additionally cause injury to people and their financial advantages. The contamination may come from an assortment of sources, going from point source water contamination (from a solitary release highlight) more diffuse, topography-level causes, otherwise called agricultural contamination [39]. Once in the climate, these contaminations can have both direct impacts in encompassing environments, for example, slaughtering nearby savage life or defiling drinking water, and downstream impacts, for example, no man's lands brought about by farming overflow is moved in enormous water bodies. The agricultural contamination assume an urgent part in the sum and effect of these toxins. The spread of pesticides and manures is more common in worldwide agriculture practices. Inadequate administration rehearses incorporate inadequately direct creature taking care of tasks, overgrazing, furrowing, manure, and ill-advised, unnecessary, or severely coordinated utilization of pesticides. Contaminations from agriculture incredibly influence water quality and can be found in lakes, streams, wetlands, estuaries, and groundwater [40]. Toxins from cultivating incorporate silt, supplements, microbes, pesticides, metals, and salts. Animal agriculture outsized affects poisons that enter the climate. Air contamination brought about by agriculture through land use changes and creature agriculture rehearses immense affect environmental change. Agriculture pollutant emerges from a plenty of item types and covers a wide scope of synthetic classes. In terms of agricultural frameworks, there are various emerging toxicity sorts of likely concern, including:

- Naturally created mixtures such a toxicity liberated by growths, microorganisms, and plants
- Human personal care products items like vital oils, herbal medicines, and antibacterial
- Emerging steady natural toxins, for example, fire retardants and dioxin-like mixture
- Veterinary medicines such as antibiotics and antiparasitic agents
- Hormones like synthetic and natural estrogens and androgens
- Nanomaterials
- Human medications
- Metabolites and change results of man-made synthetic compounds that are created from biological, chemical, and physical breakdown responses

Emerging contaminants will be derived from the agricultural environment via a number of routes. Veterinary medicines and their metabolites will be liberated directly to soils (creatures at field) or by implication at the point when excrement and livestock facilities are applied to agricultural land as a fertilizer. These pesticides could give more exact, controlled, and successful utilization of pesticides, and accordingly possibly diminish the general amounts of pesticide utilized. Toxicity may likewise be established in the farming climate itself. Many man-made substances that enter agriculture frameworks will be debased by substance, physical, and natural cycles. While these responses may bring about the total breakdown of a compound (for example, they convert the compound to carbon dioxide, water and methane), in numerous examples, the cycles may bring about the development of stable moderate synthetic substances, which we call debasement or change items [41]. As examined above, toxicity begins from a wide scope of sources, and in certain cases, contributions to surface waters from agricultural exercises might be immaterial contrasted with different sources. For example, agricultural contributions of toxicity from human drugs and individual consideration items into the oceanic climate are most likely less significant than different courses of section into the regular domain. These items are regularly utilized by society constantly and consequently transmitted constantly from wastewater treatment frameworks into the sea-going climate though discharge application to land will in general occur just at specific seasons. Hence, drugs and individual consideration items are probably going to happen irregularly in agricultural waterbodies. The serious groupings of drugs and individual care items shipped from rural soils to surface waters are likewise much lower than seen in sewage profluent [40]. The toxicity where agriculture is at present the dominating wellspring of surface water defilement, contrasted with nonrural sources, are the regular poisons, veterinary meds, chemicals, and change results of man-made synthetic compounds utilized in agriculture. For certain toxicity, agriculture may be a significant wellspring of infected later on. For instance, if pesticides are created in the nanostructure later on, almost certainly, these materials will be shipped to and infect surface waters. For certain toxicity, for example bio-terrorism/harm specialists, the significance of agriculture as a wellspring of defilement is hard to envisage. The expected courses of contribution of various classes of toxicity to surface waters are summed up in. The additionally gives a sign of the overall significance of rural literately contrasted with different sources. Perceive that the overall significance of the unique source may change as per provincial contrasts in for example cultivating and wastewater treatment enumerate furthermore, to contrasts in the physical substance properties of individual toxicity [42].

4.2.2.2 *Transboundary*

Transboundary contamination is the contamination that starts in a single nation and, however, can cause harm in another nation's current circumstance, by getting borders through pathways like water or air. Contamination can be moved across hundreds and even great many kilometers. The extraordinary distances that contamination can spread imply that it is not contained inside the limits of any single country. This is the reason it is called "transboundary pollution" [43]. One of the issues with transboundary contamination is that can divert contamination from a weighty producer and store it onto a country whose outflows are moderately low. Another issue with transboundary contamination identifies with the statement above. Because of the way that all things associate, the weighty contamination that is apparent in the created world likewise gets obvious in far off zones. For example, pollutants can be seen in the Arctic region that are eventually transbounded from the main land area [44]. Various pollutants that are discharged into enviornment and their effect are listed in Table 4.1.

4.3 Emerging pollutants

Emerging pollutants are elucidated as mixtures that are not currently covered by existing water-quality directive, have not been concentrated previously, and are thought to be likely dangers to natural biological systems also, human well-being and protection (Fig. 4.2). They surround a different multiple of compounds, including drugs, abuse of medications, personal care products (PCPs), steroids and chemicals, surfactants, perfluorinated compounds (PFCs), fire retardants, industrial additives and agents, and gasoline additives, as well as their transformation products (TPs) [45]. In option, three new classes must be added to the rundown of emerging pollutants: nanomaterials, 1,4-dioxane, and pool sanitization results (DBPs). The way that natural mixtures enter the climate relies upon their example of utilization and method of use (e.g., removal of civil, mechanical and agriculture squanders, discharge of drugs furthermore, inadvertent spills). Once in the climate, they can be generally conveyed eventually between their creation through to utilize and removal. Since most emerging pollutants are from human use, their outflows are an issue for a few wastewater measures, so the investigation of the destiny of the emerging pollutants in wastewater-treatment plants (WWTP) is significant. Once discharged into the climate, emerging pollutants are liable to measures (e.g., biodegradation, chemical, and photochemical degradation) that provide to their eradication. Depending on the compartment where engineered chemicals compounds are available in the climate (e.g., groundwater, surface water, and silt) or in the technosphere (e.g., WWTPs and drinking-water offices), various changes can occur, here and there producing products that can diverge in their environmental behavior and ecotoxicological profile [46]. For example, TPs of certain pollutants are frequently more determined than their corresponding parent compounds and display more prominent harmfulness (e.g., the major biodegradation result of nonylphenol ethoxylates, nonylphenol, which is a lot more tireless than the parent compound and furthermore, can mimic estrogenic properties).

TABLE 4.1 In the below table explained about pollutants such as pollutant categorizes, examples, major sources and effects on human and environment [44].

Pollutant categorizes	Examples	Major sources	Effects on human and environment
Photochemical oxidants	Ozone (0_3), peroxyacly nitrates (PANs).	Pollutant in the troposphere liberated from NOx, eruptive organic compounds like methane or CO_2, stratospheric origin. In the stratosphere, protection from the UV yet obliterated by chlorine or bromine compounds (ozone trenching). Photochemical reactions in the atmosphere require sunlight, oxides of nitrogen and hydrocarbons.	Produce haze; irritation to eyes, nose and throat; respiratory issues; obstructing of sunlight.
Sulfur dioxide	Sulfur dioxide (SO_2); sulfur trioxide (SO_3); sulfate (SO_4)	Combustion of fossil fuels, volcanos, the transformation of natural sulfur compounds. Combustion of sulphur-containing fuel e.g., coal, petroleum extraction and refining; paper manufacturing; municipal incinerating; ore smelting for metal extraction.	SO_2 has utmost malicious impacts as it harms to human and other animal lungs and is a significant precursor to acid rain; adverse effects incorporate corrosion of paint, metals and injury or demise to animals and plants.

TABLE 4.1 In the below table explained about pollutants such as pollutant categorizes, examples, major sources and effects on human and environment [44].—cont'd

Pollutant categorizes	Examples	Major sources	Effects on human and environment
Carbon monoxide	Carbon dioxide (CO_2), carbon monoxide (CO)	Combustion of coal, oil and different fuels for energy production manufacturing and transport Biomass consuming. Deficient combustion of carbon compounds, oxidation of volatile organic compounds.	CO_2 has a significant part in the green-house, produces fragile carbonic acid adding to acid rains; CO influences human wellbeing by binding to hemoglobin, which may result about asphyxia; hemoglobin has multiple times more empathy with CO when collate with that of O_2.
Nitrogen oxides	Nitrogen oxide (NO); nitrogen dioxide ($N0_2$); nitrous oxide (N_20); nitrate ($N0_3$)	Nitrogen oxidation during combustions, thunderstorms. Burning of fuels; biomass burning; by-product in the manufacturing of fertilizers.	Peroxy acetyl nitrate (PAN) and nitric acid (HNO_3); concealment of plant growth and tissue harm; cause irritation to eyes, viral infections such as nitrate structure in atmosphere hinders the visibility through in soil stimulate the plant development.
Polycyclic hydrocarbons	Methane (CH_4), butane (C_4H_{10}), ethylene (C_2H_4), benzene (C_6H_6), propane (C_3H_8)	Incomplete combustion, atmospheric chemistry, use of solvents. Evaporation from gasoline tanks, carburettors; burning of fuels; biomass; municipal lard fills; microbial activity of sewage; industrial process involving solvents	Carcinogenic effect on humans; higher concentrations are toxic to plants and animals; can convert into harmful compounds through complex chemical atmosphere; some are more reactive with sunlight and produce photochemical smog.

Continued

TABLE 4.1 In the below table explained about pollutants such as pollutant categorizes, examples, major sources and effects on human and environment [44].—cont'd

Pollutant categorizes	Examples	Major sources	Effects on human and environment
		changes that occur in.	
Mercury (Hg)	HgO, mercury methyl	Combustion of carbons, ores refining, incinerators, natural. Greenhouse gas (GHG) emission scenarios under the intergovernmental panel on climate change.	The inhalation of elemental mercury vapors can cause neurological and behavioral disorders, such as tremors, emotional instability, insomnia, memory loss, neuromuscular changes and headaches. They can also harm the kidneys and thyroid. High exposures have also led to deaths.
Lead (Pb)	Pb	Combustion of gas with lead, incinerators, foundries and lead transformation.	Disruption of the biosynthesis of hemoglobin and anemia, kidney damage, Miscarriages and Disruption of nervous systems.
Organic carbon	Chlorofluoro carbon (CFCs), formaldehyde (CH_2O), methylene chloride (CH_2Cl_2), trichloro ethylene (C_2HCl_3).	Incomplete combustion of fossil; combustibles or biomass, emissions by plants, oxidation of gaseous organic compounds. Aerosol sprays; foam and plastics for making disposable fast-food containers; refrigeration	CFCs cause reduction in stratospheric ozone that allows greater penetration of ultraviolet light at earth surface; intensified UV radiations cause skin cancer and can have lethal effects on various; life forms.

TABLE 4.1 In the below table explained about pollutants such as pollutant categorizes, examples, major sources and effects on human and environment [44].—cont'd

Pollutant categorizes	Examples	Major sources	Effects on human and environment
Suspended particulate matter	Dust, soil, sulfate salts, heavy metal salts, fire particles of carbon (soot), silica.	Fuel combustion; building constructions; mining; thermal power stations; stone crushing; industrial processes; forest fires; refuse incineration.	Have chronic effects on respiratory system; deposition on the surface of green leaves thus interfering with absorption of $C0_2$ and release of 0_2; blocking of sunlight; particles size that range between 0.1 and 10um, cause greatest lung damage.
Liquid droplets	Sulfuric acid (H_2S0_4), nitric acid (HNO_3) —oil, pesticides, e.g., ddt and malathion.	Agriculture pesticides; fumigation; oil refineries; reactions of pollutants in the atmosphere.	Contribute to acid rains; corrosion; damage to various life forms.

Emerging pollutants can arrive at the climate by being moved and conveyed by means of various courses. The physico-chemical properties of chemicals substances (e.g., water dissolvability, fume pressing factor, and extremity) determine their conduct in the climate. The significant wellsprings of earth emerging pollutants are principally waste water treatment plants effluents and secondarily terrestrial run-offs (from rooftops, asphalts, streets, and rural land), including environmental deposition [47]. Veterinary drugs utilized for treatment and anticipation of illnesses in cultivating are intentionally brought into the climate when fluid compost is showered on rural field. At that point, veterinary drugs and their metabolites are inclined to debase soil and groundwater. The vehicle of veterinary drugs to groundwater can be through draining or run-off from animal slurries, while sorption of the medication in the dirt can defer its dissemination. Human pharmaceuticals enter amphibian frameworks after ingestion and ensuing discharge as the nonused parent compounds or as metabolites through wastewater treatment plants. The incorporation of pharmaceuticals and their human metabolites is vulnerable to flow through wastewater treatment plants; they would then be able to enter to waterways or streams. They can reach groundwater subsequent to filtering. Also, pharmaceuticals can arrive at surface waters by run-off from fields treated with digested sediment. The major sources

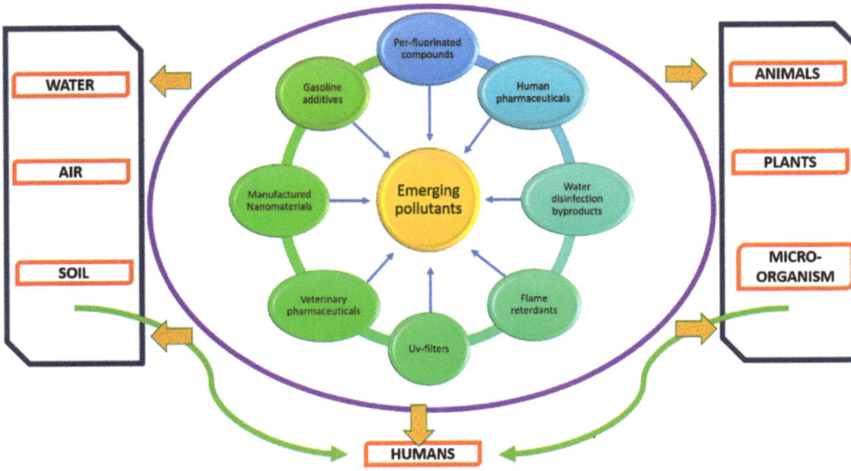

FIGURE 4.2 Categorization of emerging pollutants impact on soil, water, air, animals, plants, and microorganism and adversely effect on human.

of emerging pollutants are pharmaceutical industry, personal care products, pesticides, per-fluorinated compounds, water disinfection byproducts, and heavy metals [45].

The significant causes of emerging pollutants are drugs, individual consideration items, and manure, which could not stay away from in regular daily existence, yet the very least convergence of this emerging pollutants cause eco-friendly to human health impacts for people just as domesticated animals. The pretreatment essentially includes extraction, filtration, and derivatization. After pretreatment, it can investigated utilizing by the purification process. Emerging impurities cannot viably treat by customary wastewater treatment strategies. It requires extra treatment all together to totally take out the pollutants since insignificant focus brings adverse impact [45,48]. Treatment strategies like adsorption, layer innovation, progressed oxidation measure, and built wetlands have been talked about. Relatively, adsorption stands apart because of its effortlessness, modest, and straightforwardness to work, but it needs to pick fitting adsorbents. Other headway of treatment of emerging contaminants substances are hybrid treatment like coupling the photo-catalysis and biodegradation, improvement of biological treatment, and anaerobic layer bioreactor. Not many of the disadvantages of adsorption can be corrected with execution of late headways like nano adsorbents, cross breed framework, and adjusted adsorbents. Full disposal of emerging pollutants might be fitting, and the utilization of developments in water treatment could be essential to human well-being and the presence of emerging pollutants in regular biological system [49].

4.4 Water purification technologies

To eradicate above emerging pollutants for the treatment of wastewater, one of the best technologies is the water purification. The pervasiveness of emerging pollutants in squander water has made analyze about the public health and the environment impacts of these contaminations. Therefore, such debasement substances end up in the environment and could make a wide cluster of results each human being along with resistant bacteria, marine life organic entity undermining, neurotoxic impacts, endocrine interruption, and tumors. Water purification is the proceeding toward eliminating infelicitous chemicals, biological contaminants, suspended solids, and gases from water. The objective is to create water fit for explicit purposes. Most water is purified and disinfected for human utilization (drinking water); however, water purification may likewise be completed for an assortment of different purposes, including clinical, pharmacological, substance, and industrial applications. Regular wastewater treatment approaches incorporate a combination of physical, chemical, and biologic systems and sewage expulsion activities to dispense with insoluble pollutants and soluble pollutants [48]. To accomplish treatment, various techniques have been inaugurated to extract emerging pollutants; various of them are adsorption, membrane technology, advanced oxidation processes, biological methods, and wetlands. The performance of eliminating emerging contaminants in wastewater relied around upon its hydrophobic nature, perceptible of any hydrophobic nature, and higher qualities recommend greater wettability. Numerous investigations have been completed for the cleaning of wastewater also, to safeguard new water sources. Numerous methods have been shown as effective to kill pollutants from wastewater including adsorption, chemical precipitation, ion-exchange, conventional coagulation, electrolysis, and membrane technology. On the other side, a few potential advances have been created to eliminate heavy metals and natural pollutants for in-situ remediation of modern or home-grown wastewater [50].

4.5 Conclusion

In conclusion, the present chapter shed light on water scarcity, water pollutants, source of pollutants, emerging pollutants, and various advanced water purification technologies. Broadly, sources of water pollution can be divided into point sources and nonpoint sources. Point sources of contamination are released where toxins accumulated by a sewage waste, pharmaceutical industry, personal care products, oil pollution, etc. are hazardous waste for human health. Point sources can be effectively controlled, while diffuse sources present extraordinary trouble as far as assortment and control. On form of advanced treatment in filtration is used to remove the very small particles. Other side, non-point sources are widely dispersed in the environment and is associated with a variety of human activities, they include pollution like agriculture and transboundary pollution. These activities produce pollutants such as contaminated waters, soil erosion, sedimentation, heavy metals, and creeping of industrial discharge, that has a measured impact on adjacent countries. The anticipation of non-point sources by the use of pest control agents, pathogens, parasites etc.

Emerging pollutants are synthetic and natural occurring chemicals that are commonly monitored in the environment but have high potential to enter to environment to cause suspected adverse effect to aquatic and human health effects. Based on current investigation and studies, the emerging contaminants has mutagenicity, teratogenicity, and carcinogenicity to human being and animals. The discharge of the emerging pollutants such as antibiotics, analgesics, anti-inflammatory drugs, psychiatric drugs, steroids and hormones, utterly affects health of any living organisms. Thus, there is an emergency for an efficient water purification process to separate undesired chemicals compounds, organic and inorganic materials, and the biological contaminants from the water. Various water purification processes such as adsorption, membrane technology, chemical precipitation, ion-exchange, coagulation, and electrolysis are known to be promising techniques for water purification.

Acknowledgments

SKN thank DST-Government of India for NANOMISSION PROJECT (SR/NM/NT-1073/2016) and DST-INSPIRE (IFA12-CH-84) for financial support. SKN also acknowledges "Talent Attraction Program funded by the Community of Madrid Spain (Ref. 2017-T1/AMB5610)". Authors also thank DST-SERB (EEQ/2017/000417) for research grant. The authors acknowledge CNMS, Jain University, Bangalore, India for providing infrastructure and Nanomission Project "SR/NM/NS-20/2014" for characterization facility.

References

[1] M.K. Uddin, A review on the adsorption of heavy metals by clay minerals, with special focus on the past decade, Chem. Eng. J. 308 (2017) 438–462.

[2] P. T.C, S.K. Sharma, M. Kennedy, Nanoparticles in household level water treatment: an overview, Separ. Purif. Technol. 199 (2018) 260–270.

[3] K. Buruga, H. Song, J. Shang, N. Bolan, T.K. Jagannathan, K.-H. Kim, A review on functional polymer-clay based nanocomposite membranes for treatment of water, J. Hazard Mater. 379 (2019).
[4] S. Madhav, A. Ahamad, A.K. Singh, J. Kushawaha, J.S. Chauhan, S. Sharma, P. Singh, Water pollutants: sources and impact on the environment and human health, in: Sensors in Water Pollutants Monitoring: Role of Material, 2020, pp. 43–62.
[5] J. Ge, S. Yoon, N. Choi, Application of fly ash as an adsorbent for removal of air and water pollutants, Appl. Sci. 8 (7) (2018).
[6] K.M. Lee, C.W. Lai, K.S. Ngai, J.C. Juan, Recent developments of zinc oxide based photocatalyst in water treatment technology: a review, Water Res. 88 (2016) 428–448.
[7] M. Haji Gholizadeh, A.M. Melesse, L. Reddi, Water quality assessment and apportionment of pollution sources using APCS-MLR and PMF receptor modeling techniques in three major rivers of South Florida, Sci. Total Environ. 566–567 (2016) 1552–1567.
[8] A.T. Markad, A.T. Landge, B.B. Nayak, A.B. Inamdar, A.K. Mishra, A multivariate statistical approach for the evaluation of spatial and temporal dynamics of surface water quality from the small reservoir located in the drought-prone area of South-West India: a case study of Tiru reservoir (India), Environ. Sci. Pollut. Control Ser. 28 (24) (2021) 31013–31031.
[9] M.M. Mekonnen, A.Y. Hoekstra, Global anthropogenic phosphorus loads to freshwater and associated grey water footprints and water pollution levels: a high-resolution global study, Water Resour. Res. 54 (1) (2018) 345–358.
[10] E.V. Lopez-Ponnada, T.J. Lynn, M. Peterson, S.J. Ergas, J.R. Mihelcic, Application of denitrifying wood chip bioreactors for management of residential non-point sources of nitrogen, J. Biol. Eng. 11 (1) (2017).
[11] H. Merrikhpour, S. Mahdavi, Heavy-metal contamination and solid-phase fractionation in street dust, Arch. Environ. Occup. Health 72 (5) (2016) 279–288.
[12] L. Lamastra, N.A. Suciu, M. Trevisan, Sewage sludge for sustainable agriculture: contaminants' contents and potential use as fertilizer, Chem. Biol. Technol. Agricul. 5 (1) (2018).
[13] J. Singh, P. Yadav, A.K. Pal, V. Mishra, Water pollutants: origin and status, in: Sensors in Water Pollutants Monitoring: Role of Material, 2020, pp. 5–20.
[14] T. Chen, C. Lei, B. Yan, L.-l. Li, Analysis of heavy metals fixation and associated energy consumption during sewage sludge combustion: bench scale and pilot test, J. Clean. Prod. 229 (2019) 1243–1250.
[15] L. Cizmas, V.K. Sharma, C.M. Gray, T.J. McDonald, Pharmaceuticals and personal care products in waters: occurrence, toxicity, and risk, Environ. Chem. Lett. 13 (4) (2015) 381–394.
[16] G. Crini, E. Lichtfouse, Advantages and disadvantages of techniques used for wastewater treatment, Environ. Chem. Lett. 17 (1) (2018) 145–155.
[17] P. Chakraborty, S.N. Khuman, S. Selvaraj, S. Sampath, N.L. Devi, J.J. Bang, A. Katsoyiannis, Polychlorinated biphenyls and organochlorine pesticides in river brahmaputra from the outer himalayan range and river hooghly emptying into the bay of bengal: occurrence, sources and ecotoxicological risk assessment, Environ. Pollut. 219 (2016) 998–1006.
[18] N.H. Tran, M. Reinhard, K.Y.-H. Gin, Occurrence and fate of emerging contaminants in municipal wastewater treatment plants from different geographical regions-a review, Water Res. 133 (2018) 182–207.
[19] B.S. Rathi, P.S. Kumar, P.-L. Show, A review on effective removal of emerging contaminants from aquatic systems: current trends and scope for further research, J. Hazard Mater. (2021) 409.
[20] E. Kim, C. Jung, J. Han, N. Her, C.M. Park, M. Jang, A. Son, Y. Yoon, Sorptive removal of selected emerging contaminants using biochar in aqueous solution, J. Ind. Eng. Chem. 36 (2016) 364–371.

[21] J. Wang, S. Wang, Removal of pharmaceuticals and personal care products (PPCPs) from wastewater: a review, J. Environ. Manag. 182 (2016) 620–640.
[22] K. Lei, F. Qiao, Q. Liu, Z. Wei, H. Qi, S. Cui, X. Yue, Y. Deng, L. An, Microplastics releasing from personal care and cosmetic products in China, Mar. Pollut. Bull. 123 (1–2) (2017) 122–126.
[23] A.J. Ebele, M. Abou-Elwafa Abdallah, S. Harrad, Pharmaceuticals and personal care products (PPCPs) in the freshwater aquatic environment, Emerg. Contamin. 3 (1) (2017) 1–16.
[24] A. Álvarez-Martín, M.S. Rodríguez-Cruz, M.S. Andrades, M.J. Sánchez-Martín, Application of a biosorbent to soil: a potential method for controlling water pollution by pesticides, Environ. Sci. Pollut. Control Ser. 23 (9) (2016) 9192–9203.
[25] H.A. Ghazy, M.A.S. Abdel-Razek, A.F. El Nahas, S. Mahmoud, Assessment of complex water pollution with heavy metals and Pyrethroid pesticides on transcript levels of metallothionein and immune related genes, Fish Shellfish Immunol. 68 (2017) 318–326.
[26] P. Fierro, C. Valdovinos, C. Lara, G.S. Saldías, Influence of intensive agriculture on benthic macroinvertebrate assemblages and water quality in the aconcagua river basin (Central Chile), Water 13 (4) (2021).
[27] U.V.K. Kumar, Coastal water pollutants, in: Water Encyclopedia, 2005.
[28] S. Diwakar, K. Rajkumar, Preparation of super hydrophobic loofah sponge for fast and efficient separation of oil from seawater, Mater. Today Proc. 5 (6) (2018) 14367–14374.
[29] U. Baig, M. Faizan, M. Sajid, Multifunctional membranes with super-wetting characteristics for oil-water separation and removal of hazardous environmental pollutants from water: a review, Adv. Colloid Interface Sci. (2020) 285.
[30] F. Gao, L. Zhao, F. Shaffer, R. Golshan, M. Boufadel, T. King, B. Robinson, K. Lee, Prediction of oil droplet movement and size distribution: Lagrangian method and vdrop-J model, Int. Oil Spill Conf. Proc. 2017 (1) (2017) 1194–1211.
[31] M. Mityagina, O. Lavrova, Satellite survey of inner seas: oil pollution in the black and caspian seas, Rem. Sens. 8 (10) (2016).
[32] S.M. El-Bahi, A. Sroor, G.Y. Mohamed, N.S. El-Gendy, Radiological impact of natural radioactivity in Egyptian phosphate rocks, phosphogypsum and phosphate fertilizers, Appl. Radiat. Isot. 123 (2017) 121–127.
[33] (a) A. Bo, S. Sarina, H. Liu, Z. Zheng, Q. Xiao, Y. Gu, G.A. Ayoko, H. Zhu, Efficient removal of cationic and anionic radioactive pollutants from water using hydrotalcite-based getters, ACS Appl. Mater. Interfaces 8 (25) (2016) 16503–16510; (b) S.K. Verma, Sources of nuclear pollutants and their controls, in: Pollutants from Energy Sources, 2019, pp. 139–147.
[34] P. Samanta, P. Chandra, S. Dutta, A.V. Desai, S.K. Ghosh, Chemically stable ionic viologen-organic network: an efficient scavenger of toxic oxo-anions from water, Chem. Sci. 9 (40) (2018) 7874–7881.
[35] B. Salbu, V. Kashparov, O.C. Lind, R. Garcia-Tenorio, M.P. Johansen, D.P. Child, P. Roos, C. Sancho, Challenges associated with the behaviour of radioactive particles in the environment, J. Environ. Radioact. 186 (2018) 101–115.
[36] L. Yuan, T. Sinshaw, K.J. Forshay, Review of watershed-scale water quality and nonpoint source pollution models, Geosciences 10 (1) (2020).
[37] G. Zheng, X. Cui, S. Li, F. Zhu, Y. Chen, S. Li, Research on non-point source pollution and prevention countermeasures of river basin based on remote sensing and GIS, Arabian J. Geosci. 14 (6) (2021).
[38] M. He, X. Wang, F. Xu, G. Dong, R. Liu, Spatial-temporal characteristics of phosphorus in non-point source pollution with grid-based export coefficient model and geographical information system, Water Sci. Technol. 71 (11) (2015) 1709–1717.

[39] E.K. Zimmerman, J.C. Tyndall, L.A. Schulte, Using spatially targeted conservation to evaluate nitrogen reduction and economic opportunities for best management practice placement in agricultural landscapes, Environ. Manag. 64 (3) (2019) 313–328.

[40] B.E. Young, N. Dodge, P.D. Hunt, M. Ormes, M.D. Schlesinger, H.Y. Shaw, Using citizen science data to support conservation in environmental regulatory contexts, Biol. Conserv. 237 (2019) 57–62.

[41] M.A. Atwell, M.N. Wuddivira, M. Wilson, Sustainable management of tropical small island ecosystems for the optimization of soil natural capital and ecosystem services: a case of a Caribbean soil ecosystem—aripo savannas Trinidad, J. Soils Sediments 18 (4) (2017) 1654–1667.

[42] C. O'Bannon, J. Carr, D.A. Seekell, P. D'Odorico, Globalization of agricultural pollution due to international trade, Hydrol. Earth Syst. Sci. 18 (2) (2014) 503–510.

[43] M. Yarime, A. Li, Facilitating international cooperation on air pollution in east asia: fragmentation of the epistemic communities, Global Policy 9 (2018) 35–41.

[44] (a) A. Farahat, Air pollution in the Arabian Peninsula (Saudi Arabia, the United Arab Emirates, Kuwait, Qatar, Bahrain, and Oman): causes, effects, and aerosol categorization, Arabian J. Geosci. 9 (3) (2016); (b) M.-E. Héroux, H.R. Anderson, R. Atkinson, B. Brunekreef, A. Cohen, F. Forastiere, F. Hurley, K. Katsouyanni, D. Krewski, M. Krzyzanowski, N. Künzli, I. Mills, X. Querol, B. Ostro, H. Walton, Quantifying the health impacts of ambient air pollutants: recommendations of a WHO/Europe project, Int. J. Publ. Health 60 (5) (2015) 619–627; (c) L. Su, S.R. Swanson, The effect of destination social responsibility on tourist environmentally responsible behavior: compared analysis of first-time and repeat tourists, Tourism Manag. 60 (2017) 308–321.

[45] A.J. Hoffman, P.D. Jennings, Institutional theory and the natural environment, Organ. Environ. 28 (1) (2015) 8–31.

[46] Y. Zhou, J. Meng, M. Zhang, S. Chen, B. He, H. Zhao, Q. Li, S. Zhang, T. Wang, Which type of pollutants need to be controlled with priority in wastewater treatment plants: traditional or emerging pollutants? Environ. Int. (2019) 131.

[47] Y. Patiño, E. Díaz, S. Ordóñez, E. Gallegos-Suarez, A. Guerrero-Ruiz, I. Rodríguez-Ramos, Adsorption of emerging pollutants on functionalized multiwall carbon nanotubes, Chemosphere 136 (2015) 174–180.

[48] K. Pozo, S.K. Sarkar, V.H. Estellano, S. Mitra, O. Audi, P. Kukucka, P. Přibylová, J. Klánová, S. Corsolini, Passive air sampling of persistent organic pollutants (POPs) and emerging compounds in Kolkata megacity and rural mangrove wetland Sundarban in India: an approach to regional monitoring, Chemosphere 168 (2017) 1430–1438.

[49] M. Tiwari, S.K. Sahu, G.G. Pandit, Distribution and ecotoxicological concerns of persistent organic pollutants in sediment from creek ecosystem, J. Environ. Sci. Health, Part B 51 (9) (2016) 616–621.

[50] S. Bolisetty, M. Peydayesh, R. Mezzenga, Sustainable technologies for water purification from heavy metals: review and analysis, Chem. Soc. Rev. 48 (2) (2019) 463–487.

CHAPTER

5

A brief overview on advances in water treatment process

Amit Kumar[1], *Leena V. Bora*[2], *Nikita P. Chokshi*[1], *Gajendra Kenil*[1], *and Mansuri Zaid*[1]

[1] Department of Chemical Engineering, Institute of Technology, Nirma University, Ahmedabad, Gujarat, India; [2] School Engineering, Nirma University, Ahmedabad, Gujarat, India

5.1 Introduction

As per a recent report of the United Nations environmental program, around 33% of the world population currently stays in a grim situation with limited water, and by 2025 this percentage shall increase to 66% [1]. Presently, around four billion people are experiencing a shortage of clean and sanitized water supply and dying from various severe waterborne diseases annually [2]. The number is expected to grow further in near future. Studies [3–6] suggest that natural water cycle is polluted to a great extent due to excess discharge of contaminants and micropollutants into the nature cycle. On the other hand, increasing environmental pollution concerns and more stringent regulation standards imposed by the government of various countries have challenged various industries in seeking appropriate wastewater treatment technologies. To harvest more feasible water resources and to fulfill the increasing demand of clean water, several practical strategies and solutions are being implemented. The wastewater treatment processes depend on the source of pollution. Waste from domestic and sewer, industrial, agricultural waste is the main anthropogenic source of wastewater. Municipal and agriculture wastewater can be treated and reused for agricultural and industrial use [6]. Treated water from plants may be associated with the presence of suspended solids, colifins, soluble refractories, and other organic compounds that are further tedious and expensive to treat [7]. Industrial waste contains various types of heavy metals like arsenic, cadmium, lead, pyridine,

3D Printing Technology for Water Treatment Applications
https://doi.org/10.1016/B978-0-323-99861-1.00011-4

and phenols, which can cause carcinogenic diseases and several health ailments. Such untreated wastewater disposed into the river and sea have severely threatened the ecology of the aquatic life [8].

Emphasis has led to reducing the waste thrown out in the environment. One of the most effective ways is to reduce the waste produced in the industrial unit itself. One of the big problems associated with the treatment of wastewater from industries is scarcity of pure water around the globe which makes it important for us to minimize the effect of anthropogenic activities, on the water bodies. One such step is to reuse the water coming from various sources like industries, households, etc.

The various methods are listed in detail and a reviewed study is done on the same. The following wastewater treatment processes have been discussed in this chapter.

(i) Suspended solid removal (ii) Adsorption (ii) Ion exchange (iii) Reverse osmosis (iv) Electrodialysis (v) Chemical oxidation processes (vi) Nutrient removal (vii) Moving bed biofilm reactor (viii) Solar desalination etc.

5.2 Suspended solid removal

The solid present in the wastewater is demarcated as a total suspended solid particle (TSS) and dissolved solid based on the size of the solid particle. The size greater than 2 microns is considered as a total suspended particle (TSS) and below the size of 2 microns is dissolved solid. These solid suspended particles may be inorganic or organic material, though depending upon the types of wastewater algae and bacterial may also contribute to it. Due to the decay of plants, animals, and algae, organic material breaks away into small particles and are present as suspended solids in the water column. More the number of suspended particles in the water, the less clear the water will be; so, clarity of the water is measured from the amount of total suspended solids present in the water. Heavier suspended solid particles such as sand and gravel settled down at the bottom at still or low water flow conditions. Those solids which do not settle out into the bottom are called colloidal or not-settleable solids. These particles are too small and light to settle at the bottom.

The settleable solids can vary from fine silt and clay to larger sand and gravel and are known as bedload 37 or bedded sediments. When these sediments are moved to the bottom of the water level by strong flow, then it is called bedload transport. Movement of the settleable solid particles may be hindered with or without rejoining [9,10].

The following section discusses the different ways to remove the suspended particles from water.

1. Sedimentation
2. Centrifugation
3. Coagulation and Flocculation

5.2.1 Sedimentation

Sedimentation is a natural phenomenon by which the heavier particles settled down at the bottom of the liquid surface due to net negative buoyant force. It is part of the primary treatment process in sewage treatments and industrial wastewater treatment processes. It is one of the most commonly used methods in domestic and industrial sewage treatment plants especially when sewage contains a high percentage of "settleable" suspended solids in the water body. To achieve high-efficiency primary sedimentation followed by secondary sedimentation, tanks may be used in the wastewater treatment process (Fig. 5.1) [11,12].

Different types of clarifiers like stealable, triangular, or circular are commonly used. The efficiency of clarifiers depends on several factors like wastewater characteristics, depth of classifiers, detention period, bottom surface area, overflow rate, temperature, particle size, inlet, outlet design, etc. To separate small particles size (<2 microns) in suspension, particles are allowed to flocculate and buoyed to the surface of a liquid by adhering to the surface of many small air bubbles. The accumulated air bubbles with attached particles are then removed from the surface of the liquid by skimming.

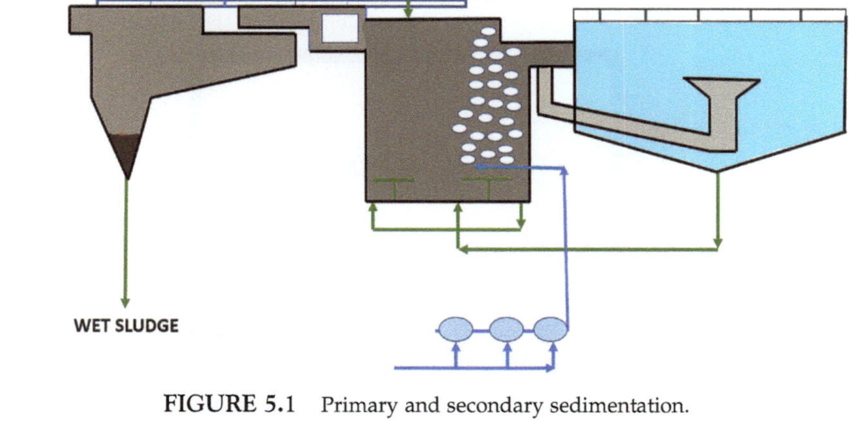

FIGURE 5.1 Primary and secondary sedimentation.

Theory: [13,63]
The settling particle follows Stokes' Law.

$$v = g_c \frac{\Delta\rho(d)^2}{18} \tag{5.1}$$

Where v is the velocity of the particle in a continuous phase in cm/s, g_c is the acceleration due to gravity in cm/s^2, d is the diameter of the dispersed particle in cm, $\Delta\rho$ difference in density of the dispersed and continuous phase in g/cm^3, and μ is the dynamic viscosity of the continuous phase in g/cm.s.

Gravity settling tanks are generally used as a primary treatment in which the natural phenomenon of sedimentation is used. The resultant velocity of the horizontal and vertical velocity defines the trajectory of particles as shown in Fig. 5.2 [14].

5.2.2 Centrifugation

The continuous separation of solid particles with efficiency is accomplished by centrifugation of sludge mixtures. All centrifuges work on the principle of pressure drop. Desanding hydrocyclones or decanters are mainly used for dewatering [15–17].

When the sludge mixture is allowed to rotate into a centrifuge, two apposite forces—outward centrifugal force and inward drag force—act on solid particles. A mixture of liquids and solids is allowed to enter in the cyclone through the volute inlet. The operative feed pressure is maintained at the inlet to fix the desired flow rate. Accelerating centrifugal force separates heavier particles as underflow based on size, shape, and density. Due to the angular acceleration of the flow pattern, centrifugal force acts on the mixture to spin in a radial vortex down the length of the cone. Thus, heavier particles are discharged continuously through the apex creating the underflow stream. Along the axis of the cyclone,

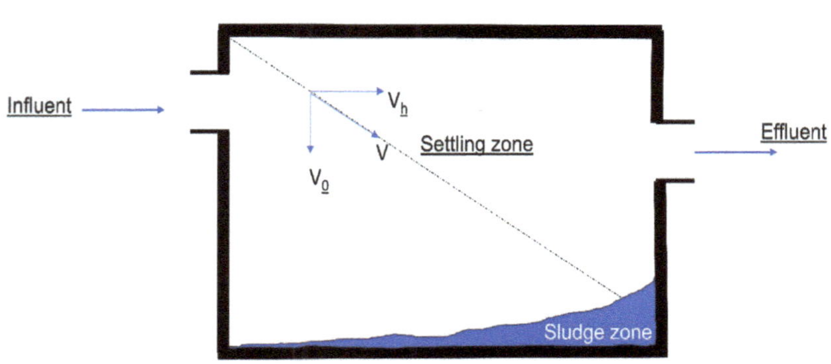

FIGURE 5.2 Schematic of the ideal settling tank.

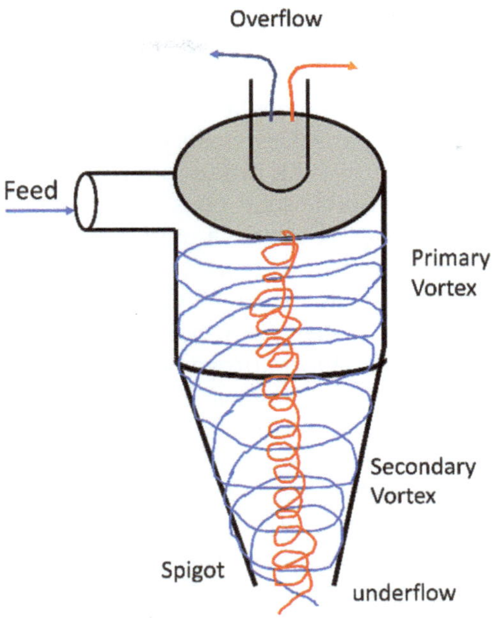

FIGURE 5.3 Cyclone separator in operation.

because of cone convergence, high pressure to the low-pressure zone from bottom to top the liquid flow is reversed hence, drag force acts on lighter particles which carried out upward through the vortex finder as overflow. The solids are collected as basic purges at the exit through the apex into the accumulation chamber. Fig. 5.3 shows a schematic of a cyclone separator in operation.

5.2.3 Coagulation/flocculation

The coagulation/flocculation process is a simple and effective process widely used in the treatment of wastewater. In this process, the physical state of dissolved or suspended particles is altered by the addition of chemicals which further facilitate flotation or sedimentation. The terms coagulation and flocculation are ambiguous and used interchangeably by many. However, flocculation is the process in which the particles are destabilized by making them larger aggregates, so that they settle down at the bottom, whereas the term coagulation means giving rise to aggregations by destabilizing the suspension [18]. The addition of flocculant aids may improve the coagulation or flocculation process such as the formation of stronger and denser flocs, an increase of flocs settling velocity, etc. Many researchers [19–21] studied the characteristics of suitable aids with respect to pH range, types of coagulants, physical and chemical properties of solution, etc. In recent years, attention is being paid to developing a sustainable and viable process through the flocculation/coagulation method; thus focus has been shifted toward plant-based materials as a flocculant/coagulant. Plant-based materials are nontoxic, biodegradable, easy to handle, and most importantly environmentally friendly, which takes advantage of their inorganic counterparts. Despite its potential uses, there are many obstacles to commercializing the product, for example, bulk processing, mass plantation, and favorable conditions like moderate price and high-volume market, existing supportive regulations are required. Investigations are going on to improve extraction and purification techniques for active coagulant agents. Several hybrid materials such as PASiC, PACS, starch-g-PAM, starch-chitosan, polymers having cationic cross-linking have shown remarkable improvement as coagulant/flocculation in the wastewater treatment process. Research shows that by the combination of the effective functional groups into the original material like hydroxylation-polymerization, physical blending at elevated temperature, conditioners like various salts along with acid, copolymerization and chemical grafting/cross-linking, copolymerization, immobilization, or composite polymerization significantly enhance the ability of aggregation of the material. Although promising at the lab scale its economic viability and performance efficiency in the treatment of real industrial water must be studied.

5.3 Adsorption

In the adsorption process, there is an accumulation of a substance at the interface between two phases as shown in Fig. 5.4. *Adsorbate* is a substance that accumulates at the interface and the solid on which adsorption takes place is an *adsorbent*. Basically, there are two types of adsorption: chemical sorption and physical sorption. Chemical sorption is generally irreversible and physical adsorption is generally reversible, as there are weak van der Waals forces between adsorbent and adsorbate [22]. In adsorption generally, a fluid (water in this case) and the adsorbent in the solid phase is involved. The dissolved impurities are transferred from liquid to the surface of the solid and hence the water is cleaned. Adsorption is widely used for water treatment due to the less cost, more efficiency, ease of implementation and ease of operation, availability of different solids as an adsorbent material, and chance to recover both adsorbent and adsorbate [23]. The adsorption process is affected by various parameters like temperature, the nature of the adsorbent, the concentration of pollutants, particle size of adsorbent, pH of the solution, etc. When contaminated water is in contact with a solid adsorbent, the pollutants stick to the solid surface of the adsorbent and a stage of equilibrium is established. At this point, the concentration of pollutants

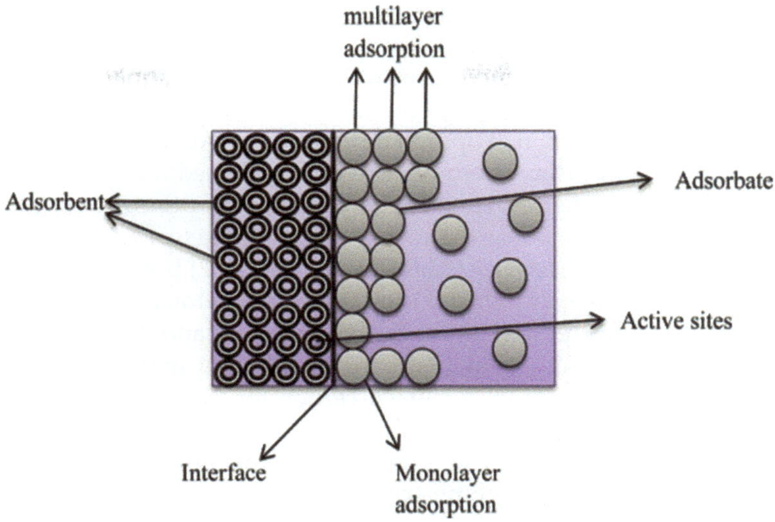

FIGURE 5.4 Basic process of adsorption. *From N. K. Soliman, A. F. Moustafa, Industrial solid waste for heavy metals adsorption features and challenges; a review. J. Mater. Res. Technol. 9 (5) (2020) 10235–10253.*

adsorbed and, in the water, become constant. The adsorption isotherms are basically the relation between the equilibrium quantity of pollutants adsorbed and in the water at a given temperature [24]. Several relationships and equations are available to describe this relationship like the Freundlich and the Langmuir isotherms (Fig. 5.4) [25].

A number of materials have been extensively used as an adsorbent in water treatment. Some of them are activated carbon, silica gel, activated alumina, zeolites, etc. [26]. Some low-cost adsorbents like agricultural waste peels, biomass-based activated carbon, and industrial byproducts have been used for water treatment [27].

Some of the adsorbents which are used in water treatment are:

Alumina is available in the form of granules of different sizes ranging from 200 to 300 m^2/g surface area. Mohammad et al. [28] used activated alumina modified with potassium hydroxide and sodium hydroxide solutions to improve the CO_2 adsorption capacity. They studied the various parameters like amount of adsorbent, percentage of hydroxide solution, and temperature and pressure. Similarly, Bhargavi et al. [29] studied the removal of zinc and color black G using alumina nanoparticles as an adsorbent. The maximum adsorption of Zn is found near 7 pH while the maximum removal for color black G was found at 2 pH. Silica gel is a porous and crystalline granule of various sizes and it has more surface area compared to alumina [25]. Activated carbon is the oldest and most widely used adsorbent. It can be prepared from coal, biomass, agricultural waste, etc., using physical and chemical activation methods [25]. Rahman et al. [30] prepared the activated carbon from rice husk and used it for the removal of methylene blue dye from an aqueous solution. The maximum uptake of methylene blue was found near 97.15% using activated rice husk carbon at optimized conditions. Abbaszadeh et al. [31] prepared activated carbon from papaya peel and used it for the removal of lead from water. The best results were obtained at optimized conditions and 93% of lead was adsorbed at pH 5, the dosage of adsorbent was 100 mg, and contact time was 2 h. Kadirvelu et al. [32] prepared activated carbon from coconut coir pith and used it for adsorption of cadmium. The adsorption data were well fitted with Langmuir and Freundlich isotherms. Similarly, Shukla et al. [33] used the coconut shell, orange peels, and banana peels to prepare activated carbon for wastewater treatment. The activated carbon prepared from orange peels gave the best results among all prepared adsorbents. The cost of activated carbon is comparatively high and even regeneration cost is also high so recently the focus is on low-cost adsorbents. Banana peel, red mud, rice hull ash, coir pith, orange peel, sunflower stalk, cotton waste, bamboo dust carbon, etc., have been used to make the process more economical and environment friendly.

5.4 Moving bed biofilm reactor (MBBR)

Moving Bed Biofilm Reactors (MBBRs) are the secondary and tertiary process for wastewater treatment. This wastewater treatment process is used to reduce inorganic and organic components from the waste matter. It is best suited for petroleum contaminated waters, piggery wastewater treatment, pharmaceutical wastewater, pulp, and paper industry waste streams, and sustainable nutrient remediation in municipal wastewater treatment [34,35] The overall performance of the MBBRs depends upon the rate of reaction which is determined by surface area (media size, shape and specific surface area) of bio carrier media, and the fill ratio of the reactor vessel [36]. MBBRs provide an attachment surface for active microorganisms into suspended biocarriers. The attachment surface is made up of small plastic carrier elements in a continuously mixed reactor. Air mixing into the reactor provides the replacement of nutrient-rich wastewater into the active biofilm surface phase which enhances nitrogen oxidation as shown in Fig. 5.5. Air mixing helps to achieve mixing of the treated water and biocarrier media. Active biomass grows as a biofilm onto the surface of the carriers having a density less than the water. MBBR is effective to retain slowly growing microorganisms such as nitrifiers in the form of biofilm.

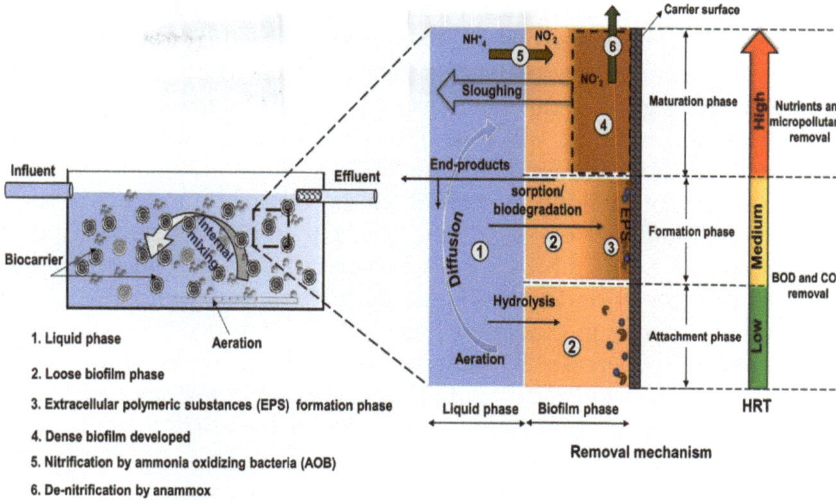

FIGURE 5.5 Principles of operation and mechanism of MBBR [34].

The primary advantage of MBBR process is it is flexible and easy to install and maintain, as it can be fitted into any size or shape of tank. It can be upgraded to any overloaded or trickling filter, activated sludge, or other processes. Most importantly, it does not require sludge recirculation as fixed bed reactor. It can operate continuously and is free from the problem of clogging. It also offers higher specific surface area for the biofilm compared to the fixed bed reactor. Further study can yield more results on the performance of the MBBR system.

The performance of MBBR for Biological Oxygen Demand (BOD) and nutrient removal efficiency has been studied by several researchers in different process combinations. The study carried out by Ref. [37] reported that for the same effective surface area of the MBBRs, the organic surface area loading rate (g COD/m^2.d) is the only main factor for the removal of organic matter and the shape and size of the carrier have negligible influence. Similarly, [38] carried out laboratory-scale experiments using different biofilms carriers and Fe(II) sulfate as coagulants. The report showed phosphorous removal efficiency of more than 75% and removal efficiency of nitrification reached more than 90% at DO 2 mg/L. The investigation carried out by Ref. [39] on MBBR filled with FLOCOR–RMP suggests an efficient option for the total nutrient removal from municipal wastewater. Experimental results on MBBR filled with FLOCOR–RMP carried by Ref. [39] demonstrate that the system can be used for almost complete organic and nutrient removal, they achieved Chemical Oxygen Demand (COD), TN removal efficiencies of 96.9, 84.6% at optimum operating conditions, respectively. The removal efficiency of organics and nitrogen from wastewater has been investigated by Ref. [40]. They took two different carrier martials, PUF and biodegradable polymer particles Polycaprolactone (PCL), for their research. They confirmed that the MBBR shows good removal efficiency for Total Organic Carbon (TOC) and ammonium removal while MBBR with biodegradable PCL carrier has a good performance for TN removal. Ref. [41] investigated on nitrification kinetics of a pilot-scale MBBR for the removal of NH_4-N and nitrite from polluted raw water. In their study, they noted a sharp drop in nitrification rate when at MBBR temperature falls to 5.0°C. He noted that NH_4N and nitrite oxidation are highly dependent on temperature. They observed an increase in oxidation rates of NH_4-N and nitrite with temperature and a decrease in biomass concentrations as temperature increased from 16.1 to 28.3°C. Based on this study, they suggested that polysaccharides (PS), phospholipids (PL), and proteins (PN) can be used as attached biomass in MBBR systems.

5.5 Solar desalination

As per a recent report of United Nations Environmental Program, around 33% of world population currently stay in grim situation with limited water and by 2025, this percent shall increase to 66%. Solar energy is a freely and abundantly available renewable source of energy. As per the findings by the International Energy Agency, it could be the highest source of electricity to the world by 2050. Sustainable water supply strategies require energy for its treatment. Solar energy shows a high technically feasible useable potential. Hence technologies need to be developed, viz., solar desalination, solar photocatalysis, solar disinfection, etc., that integrate solar energy with these processes [42].

Solar energy can be utilized in two ways to drive the desalination plants, viz., directly or indirectly. In the direct desalination system, a solar still is used which directly produces distilled water in a solar collector. On the other hand, in indirect solar desalination technology, solar energy is first harvested via solar thermal collectors which collect heat, or solar PV panels that generate electricity. This is then used for thermal or membrane distillation strategies like multieffect desalination, reverse osmosis, membrane distillation, multistage flash desalination, etc.

5.5.1 Direct solar desalination—solar still

Direct solar desalination strategy [43] includes the use of a solar still which is used in remote areas which need water supply systems having only low cost. The attractiveness of this equipment lies in its simple construction and maintenance along with less cost. The efforts and gaps

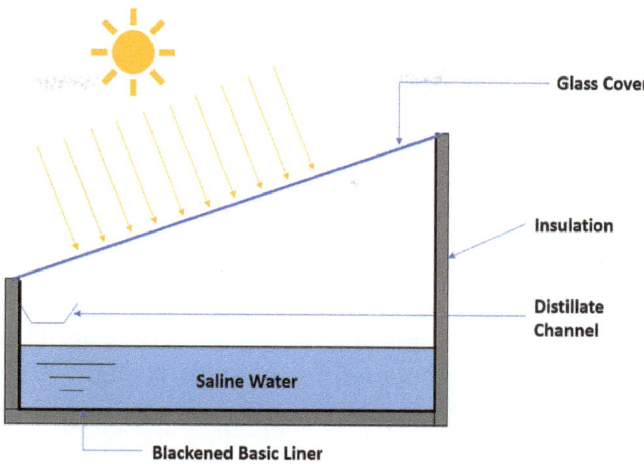

FIGURE 5.6 Schematic of a solar still. Source.

lie in enhancing its performance characteristics. It consists of a basin covered with a transparent sloping cover. The base of the basin is painted black to absorb heat and it contains the saline water. Heat gets trapped within and evaporates the water which then condenses on the inner surface of the sloping cover. This condensed water is then collected at the lower end of the sloping cover as distilled water (Fig. 5.6).

5.5.2 Indirect solar desalination

5.5.2.1 Solar humidification-dehumidification

A solar humidification-dehumidification (SHDH) technology comprises a humidifier, a dehumidifier, and a heater driven by solar energy. Since both humidification as well as dehumidification takes place in separate chambers, they can be independently designed and optimized as per conditions and parameters. The SHDH process is based on the moisture carrying property of warm air which effects into separation of pure water in the form of moisture from saline water [44]. SHDH process is used for the production of desalinated water in low capacity. The process is categorized into three classifications, viz., closed air-open water cycle, closed water-open air cycle, and open water-open air cycle. Recent advancements include the use of packed bed humidifiers and finned-tube heat exchanger in the dehumidifier.

5.5.2.2 Solar-powered multistage flash

In multistage flash (MSF) powered by solar energy, solar thermal collectors are used in conjunction with the conventional MSF plant. MSF is at present one of the most prominent thermal technologies for desalination, with a share of around 21% worldwide desalination installed capacity. In this process, the saline water is first preheated in heat exchanger by exchanging heat with vapors condensing in successive stages. This preheated brackish water is then externally heated with the help of solar thermal collectors. This then passes through each stage one after the other with reducing pressure so that at each stage, some flashing takes place. The gain output ratio (GOR) and performance of an MSF system can be increased by increasing the top brine temperature, reducing the feed temperature of the saline water, increasing the heat transfer area, and the number of stages in flashing. However, solar MSF is less competitive technologically, thermodynamically, as well as economically than solar multiple-effect distillation because, (i) the need of high TBT for better performance makes it less favorable to be used with solar heater, (ii) higher TBTs also increase the chances of fouling and scaling.

5.5.2.3 *Solar powered membrane distillation*

Solar powered membrane distillation (MD) uses difference in vapor pressures across a membrane as a driving force for water desalination. The membrane selectively allows only volatile members and water vapor to pass through a thermally driven hydrophobic membrane. The MD process is generally operated at 60–80°C. Four basic configurations have been stated on the basis of means to recover the separated vapor on the permeate side. They are: (i) direct contact membrane distillation, (ii) air gap membrane distillation, (iii) vacuum membrane distillation, and (iv) sweeping gas membrane distillation.

5.5.2.4 *Solar-powered electrodialysis*

A solar PV system drives this electrodialysis (ED) process. An electric potential generated by the PV system is applied across the cathode and anode (Fig. 5.7) of an ED stack, so that the cations and anions move toward the opposite sides, thus removing the salinity of the water.

5.6 Advanced oxidation processes

The advanced oxidation process (AOP) involves the generation of strong reactive oxygen which helps treat various types of pollutants present in water and wastewater [45,46]. AOPs majorly generate the hydroxyl radical (*OH) which in turn attacks the organic and inorganic pollutants and degrades them (Eq. 5.2) at room temperature and pressure.

$$R - H + OH\bullet \rightarrow H_2O + R\bullet \text{ ——} \quad (5.2)$$

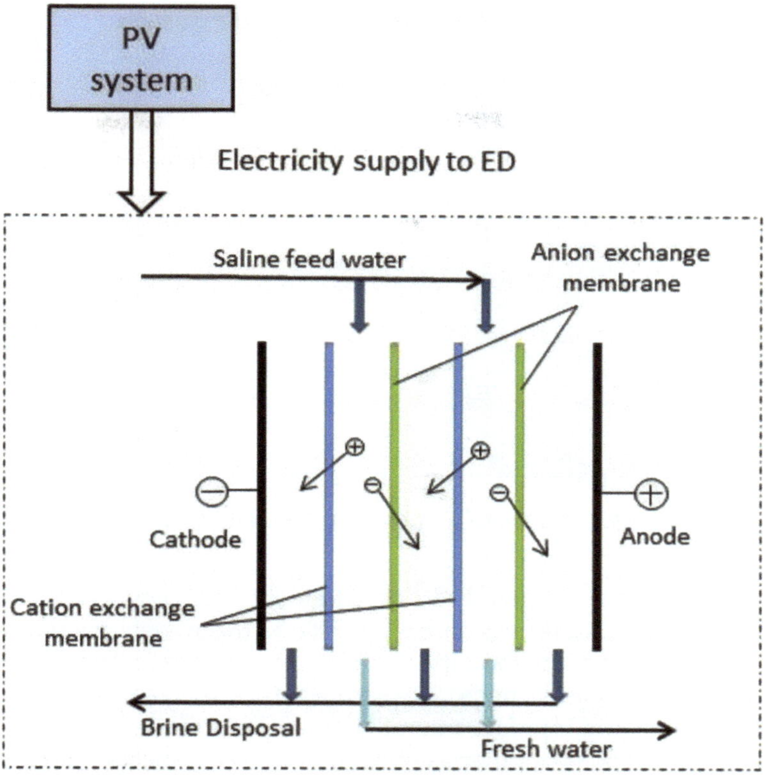

FIGURE 5.7 Schematic diagram of PV-driven ED system. *Source: Reproduced from Y. Zhang, M. Sivakumar, S. Yang, K. Enever, M. Ramezanianpour, Application of solar energy in water treatment processes: a review, Desalination, (2018) 116–145.*

Since the oxidation potential of *OH radical is 2.33 V, it makes it a quicker process than other conventional treatment processes. The most noteworthy advantage of AOPs is that they completely degrade the pollutant and don't just transfer them from one phase to another. Also, they are environmentally friendly and do not produce large amounts of hazardous sludge. The different types of AOPs are discussed as follows:

5.6.1 Sonolysis

Sonolysis is a facile AOP that is sludge-free, without the generation of any secondary pollutants [47]. Ultrasonic sound waves with a frequency range of 20 kHz–10 MHz are passed through the water/wastewater to be treated. This hydrodynamic feature produces cavitation bubbles, which upon bursting degrade the organic molecules present in the water to be treated (Fig. 5.8).

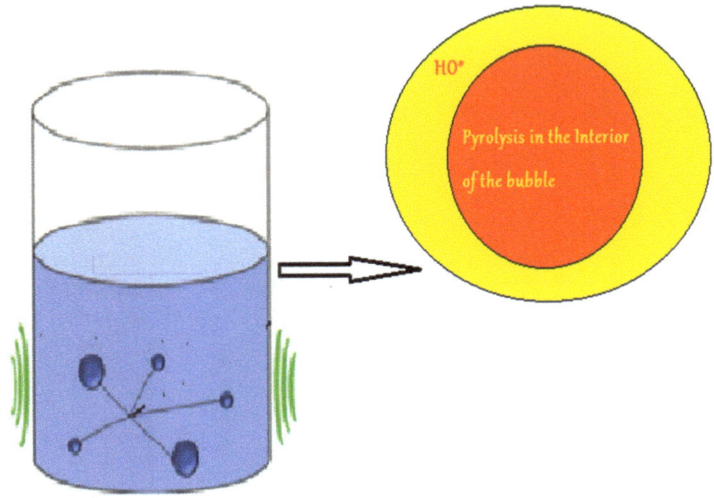

FIGURE 5.8 Schematic showing the working principle of sonolysis.

This is microscopic pyrolysis at extremely high temperatures and pressures of 500–10,000 atm and 3000–5000 K. this dissociates the water molecules leading to the formation of *OH radicals which then attack the pollutant. Secondary reactions take place when the free radicals generated at the bubble-liquid interface migrate into the bulk of the liquid.

Sonolysis is not as popular as other AOPs since it incurs high energy and gives lesser mineralization efficiency. To make it potentially more active, its intensification and/or combination with other AOPs is suggested. Hybridization of ultrasonic waves with UV accelerates the generation of ROS thereby improving the efficiency. Also, the synergistic effect of ultrasound waves with the presence of a photocatalyst sensitive to UV radiations also leads to improved bubble cavitation. Sonoozonolysis is another intensification technique wherein each O_3 molecule generates two *OH radicals. Hence the production of *OH radicals and thermolysis of ozone get enhanced by in-situ H_2O_2/HO^* generation. ROS generation and hence demineralization may be further enhanced by the addition of H_2O_2 or by the increase in system pressure.

5.6.2 Ozone-based advanced oxidation processes

Ozone, O_3, is an allotrope of oxygen that is unstable and reactive. Hence, it is almost always generated in-situ. It is generated by the application of high voltage on dry and cool gaseous oxygen (Fig. 5.9) (Eqs. 5.3 and 5.4). This process is known as *corona discharge* and is a commercially accepted technology of ozone generation.

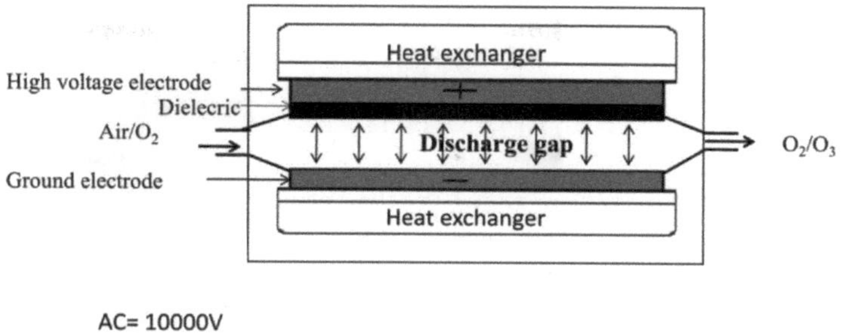

FIGURE 5.9 Schematic of corona process for generation of ozone. *Reproduced from C. V. Rekhate, J. K. Srivastava, Recent advances in ozone-based advanced oxidation processes for treatment of wastewater-a review, Chem. Eng. J. Adv. 3 (2020) 100031.*

$$O_2 + \text{electricity} \rightarrow 2O\bullet \text{ ——} \quad (5.3)$$

$$O_2 + O\bullet \rightarrow O_3 \text{ ——} \quad (5.4)$$

Its strong oxidation potential of 2.07 V makes it capable of degrading the organic pollutants via two mechanisms: (1) direct electrophilic attack caused by molecular ozone, and (2) indirect attack caused by hydroxyl radicals produced from ozone. Ozone-based AOPs are discussed further:

5.6.3 Ozonation and UV radiation (O_3/UV)

This process begins with ozone photolysis. This is then followed by the step of generation of strong hydroxyl (OH•) radicals by water reacting with O• (Eqs. 5.5, 5.6, and 5.7). UV accelerates the degradation of the pollutants by bestowing more free OH• radicals thereby enhancing the demineralization rate.

$$O_3 + UV \rightarrow O_2 + O\bullet \text{ ——} \quad (5.5)$$

$$O\bullet + H_2O \rightarrow 2OH\bullet \text{ ——} \quad (5.6)$$

$$2O\bullet + H_2 \rightarrow OH\bullet + OH\bullet \rightarrow H_2O_2 \text{ ——} \quad (5.7)$$

5.6.3.1 O_3/H_2O_2

In a combination process of O_3 and H_2O_2, commonly known as peroxone AOP, formation of strong radicals takes place by chain reaction of O_3 decomposition that is brought about by hydroperoxide $HO^-{}_2$ anion (5.8–5.12).

$$H_2O_2 \rightarrow HO^-{}_2 + H^+ \text{ ——} \quad (5.8)$$

$$HO^-{}_2 + O_3 \rightarrow HO\bullet_2 + O^{\bullet -}{}_3 \text{ ——} \quad (5.9)$$

$$O^{\bullet -}{}_3 + H^+ \rightarrow HO^{\bullet} - {}_3 \text{——} \quad (5.10)$$

$$HO_3{}^{\bullet} - \rightarrow O_2 + OH\bullet \text{——} \quad (5.11)$$

$$OH\bullet + H_2O_2 \rightarrow HO\bullet_2 + H_2O \text{——} \quad (5.12)$$

Care needs to be taken while adding the peroxide to the process as excess peroxide leads to scavenging of OH* radicals and reformation of the hydroperoxide ion ($HO\bullet_2$). Hence careful optimization needs to be carried out before using H_2O_2 (Eq. 5.13).

$$H_2O_2 \rightarrow H_2O + 12O_2 \text{——} \quad (5.13)$$

5.6.3.2 *O_3/homogeneous catalyst*

Homogeneous catalysts used in conjunction with ozone are generally transition metals like Fe^{2+}, Cu^{2+}, Cr^{2+}, Mn^{2+}, Ni^{2+}, Co^{2+}, Cd^{2+}, Ag^+, Zn^{2+}, etc. They assist and enhance the degradation of recalcitrant pollutants present in water and wastewater by increasing the reaction rate, enhancing ozone oxidation selectivity, and increasing the effectiveness and efficiency of ozone utilization. The process of catalytic ozonation follows the following steps: (i) ozone gets decomposed by the catalytic metal ions, thus producing free radicals, (ii) a chemical complex is formed between the catalyst and the organic molecule, later, this complex gets oxidized. For example, in O_3/Fe(II) process, the Fe(II) ions enhance the decomposition of ozone which in turn enhances the OH* radical formation, thereby increasing the overall efficiency of the process (Eqs. 5.14–5.16).

$$O_3 + Fe^{2+} \rightarrow FeO_2{+} + O\bullet \text{——} \quad (5.14)$$

$$H_2O + FeO^{2+} \rightarrow Fe^{3+} + OH\bullet + OH^- \text{——} \quad (5.15)$$

$$FeO^{2+} + Fe^{2+} + 2H^+ \rightarrow 2Fe^{3+} + H_2O \text{——} \quad (5.16)$$

5.6.3.3 *O_3/Heterogeneous catalysts*

Heterogeneous catalysts in ozonation include the use of metal oxides like MgO, MnO_2, Fe_2O_3, $CuFe_2O_4$, $ZnFe_2O_4$, $MgFe_2O_4$, etc., either used as it is or may be supported on a suitable substrate, like SBA, γ-alumina, zeolites, etc. Some porous materials can also be used as support like carbon nanotube, activated carbon, etc. The influencing parameters include the catalyst used, its amount, its surface area, solution pH, and target pollutant.

5.6.3.4 *Photocatalytic ozonation*

Intensification of ozonation with photocatalysis is a very promising technology study these days. Photocatalysts like titanium dioxide (TiO_2),

Fe_2O_3, $BiVO_4$, WO_3, In_2O_3, etc., are UV sensitive and get activated when UV/visible light falls on them. Upon irradiation by UV/visible light, the holes and electrons of the photocatalyst get separated and migrate to participate in redox reactions occurring with the adsorbed pollutant on the surface of the photocatalyst. This results in the generation of superoxide radical anion ($O\bullet^{-2}$) and hydroxyl radical ($OH\bullet$) (Eqs. 5.17–5. 20).

$$\text{Photocatalyst} + h\gamma \rightarrow h^{+} + e- \text{——} \tag{5.17}$$

$$O_3 + e- \rightarrow O_3{}^{\bullet -} \text{——} \tag{5.18}$$

$$O_3\bullet - + H^{+} \rightarrow HO_3\bullet - \text{——} \tag{5.19}$$

$$HO_3{}^{\bullet -} \rightarrow O_2 + OH\bullet \text{——} \tag{5.20}$$

$$O_3 + H_2O \rightarrow H_2O_2 \text{——} \tag{5.21}$$

$$H_2O_2 \rightarrow HO_2- + H^{+} \text{——} \tag{5.22}$$

The electrons that get generated by photocatalysis get trapped by ozone leading to formation of radicals (Eqs. 5.23 and 5.24) and thus the synergistic performance and intensification is caused.

$$e- + O_2 \rightarrow O_2\bullet - \text{——} \tag{5.23}$$

$$O_2\bullet - + O_3 \rightarrow O_3\bullet - + O_2 \text{——} \tag{5.24}$$

5.6.3.5 O_3/Fenton process

The Fenton process involves the use of Fe^{2+} ions in conjunction with hydrogen peroxide (H_2O_2) to generate the ROS capable of degrading organic and inorganic pollutants present in water/wastewater. Recently, several corresponding intensification options have been explored and studied, prominent among them being sono-Fenton, photo-Fenton, electro-Fenton, sono-photo-Fenton, sono-electro-Fenton, and photo-electro-Fenton. The O_3/Fenton synergism accelerates the radical generation process leading to higher treatment efficiencies. It is also suggested that Fe(II) may catalyze ozone formation as well. The reactions taking place are as follows:

$$Fe(II) + H_2O_2 \rightarrow Fe(III) + OH\bullet + OH- \text{——} \tag{5.25}$$

Fe(II) reacts with hydrogen peroxide to form the OH* radical through the transfer of electrons. The generated ROS may get scavenged to form higher complexes.

$$Fe(II) + OH\bullet \rightarrow Fe(III) + OH- \text{——} \tag{5.26}$$

Excess amounts of peroxide may scavenge OH* themselves leading to the formation of hydroperoxide ion (HO_2^-) and hydroperoxide radical ($HO\bullet_2$) (Eqs. 5.25–5.29); hence an optimum dosage is a must.

$$OH\bullet + H_2O_2 \rightarrow HO^{\bullet 2} + H_2O \text{ —— } \quad (5.27)$$

$$HO\bullet_2 + Fe(II) \rightarrow HO^{-2} + Fe(III) \text{ —— } \quad (5.28)$$

$$H_2O_2 \rightarrow HO{-}_2 + H^+ \text{ —— } \quad (5.29)$$

Ozone itself generates OH• through a series of steps:

$$O_3 + e- \rightarrow O_3\bullet - \text{ —— } \quad (5.30)$$

$$O_3\bullet - + H^+ \rightarrow HO_3\bullet - \text{ —— } \quad (5.31)$$

$$HO_3\bullet - \rightarrow O_2 + OH\bullet \text{ —— } \quad (5.32)$$

Thus the synergistic effect of ozonation with the Fenton process results in the enhanced generation of OH• radicals leading to an increase in the treatment efficiency.

5.7 Reverse osmosis

Reverse osmosis a common technique used for desalination as well as treatment of water. It uses a semipermeable membrane that rejects salt/pollutants thereby cleaning water. RO plants are very common and used extensively in biotechnology, mine and diary wastewater, pharmaceuticals, power generation, food processing, beverage industry, textiles, semiconductors, coproducing water from oil and gas production, process and boiler water, tanneries, pulp, and paper industry, among others, apart from desalination. The influencing parameters/factors include differences in the pressure across the membrane, the pore size of the membrane, charge exclusion, and physicochemical interactions between the membrane, solvent, and solute. Spirally wound and hollow fiber are the most common commercially available RO modules. Several good review articles are available in the literature [48,49].

Membrane fouling is an important limiting factor in all RO systems [50]. It is the result of clogging and/or adsorption of salt/pollutants on the surface of the membrane. Membrane fouling may be caused by (i) any microbial growth, (ii) scaling, (iii) particulate and colloidal matter forming cakes, and (iv) dissolved organic substances. Hence, before the feedwater enters the RO system, it needs to be pretreated. This prevents fouling and thus leads to prolonged membrane life [51]. This pretreatment is usually affected by fine filtration, ultrafiltration, microfiltration, chemical coagulation, acidification, etc. Pretreatment results in less volume of reject brine and higher recovery rates.

Membrane characterization helps us to have a good understanding of its properties as well as levels of molecular, microcrystalline, and colloidal characters present in it. Novel techniques are being continually developed to facilitate researchers to study the same. Some of these methods include liquid-solid equilibrium or thermoporometry, bubble gas transport, liquid-vapor equilibrium, solute rejection, value of water flux, gas-liquid equilibrium or thermoporometry, molecular weight cut-off estimation, mercury porosimetry, and microscopic methods.

The total cost of the RO system includes fixed cost as well as operating cost, viz., the cost of energy consumption as well as membrane replacement. The total water desalination cost is often cut by using RO in conjunction with other desalination techniques. Energy consumption and recovery issues may be addressed by the usage of the turbocharger, Pelton wheel, pressure exchanger, etc. These machines can transfer the energy present in the brine to high-pressure pumps used to drive the RO plant. According to a recent study, an energy recovery system may reduce the energy consumption by 30% which may further be reduced to around 5% with proper design and operation. Also, the addition of certain chemicals may accelerate the softening process to a great extent. Another strategy suggested is to increase the number of RO units in series, automate the system, and increase the size of individual RO units. These strategies would ensure higher effectiveness and efficiency, better membrane fouling, and fault control. An increasing number of stages increases water recovery and reduces specific power consumption. RO systems may also be driven by gas/steam. This reduces operating costs.

The reject brine also needs to be disposed of properly. This depends on the properties of the brine, its pretreatment before disposal, its volume, and the method used for its disposal. Currently, the conventional methods of disposal include injecting into a well, use of evaporating ponds, and direct sea discharge. Of the three methods mentioned here, the cost of the former is the highest than the latter options. However, evaporating ponds are natural and allow for resource recovery.

Continual research is carried out to design innovative RO membranes. Some of them are capable of treating highly concentrated saline water. They are capable of recovering as much as 60% of the brine, and converting it to potable water with an appropriate TDS count of about 200 ppm. They could also reduce the plant size by around 30% and cost by around 15%. RO systems have also been integrated with solar PV which helps reduce the load on fresh electricity to a great extent. Some innovative membranes are made up of cross-linked polymers and specially engineered nanoparticles. They are capable of drawing significantly more amount of water ions while repelling major amounts of contaminants like bacteria, dissolved salts, and organic materials. This also results in less pumping cost of water through the membrane since fouling and adhesion on the membrane is minimal in such cases. Novel membrane-like materials have been designed to be used, made up of polyether-urea and polyamide-urea. They are capable of reducing the microbial adhesion and hence fouling of the membrane.

5.8 Electrodialysis

Day by day the scarcity of water is increasing; so it is necessary to find different sustainable ways to reserve and treat water. There are different thermal and membrane desalination techniques adopted worldwide to treat saline water and convert it into drinkable water. Membrane processes use a membrane film as a barrier to separate the pollutants from wastewater, while in thermal processes high energy is used to vaporize the water and leave the contaminants [52]. Among them electrodialysis is widely accepted as it doesn't depend on high-pressure values for desalination, high salt removal rate, less risk of scaling, and it is widely accepted to treat the saline water with a salinity of 1000–5000 mg/L total dissolved solids [53]. There are different applications of the electrodialysis process in the field of industrial wastewater treatment, saline water treatment, and municipal wastewater treatment to remove the ionic and nonionic contaminants under the electric current [52].

5.8.1 Basic principle of electrodialysis

In the electrodialysis process, the unit consisting of a series of cation exchange membranes (CEM), and anion exchange membranes (AEM) is located between the two electrodes, which generates the concentrate and dilute solutions compartment alternatively as shown in Fig. 5.10 [54]. As the electric current flows through the system, anions from the feed solution move toward the positively charged electrode, and cations move toward the negatively charged electrode [54]. These ion exchange membranes have either positive or negative active groups fixed to their surface. Some basic characteristics of ion exchange membranes are little electric resistance, high ion-exchange capacity, higher mechanical strength, chemical resistance, and electivity [55].

Albornoz et al. [56] showed the application of electrodialysis to the university wastewater to reuse and recover water. The results showed that above 80% of all cations and 70% of all anions were removed hence the contaminants are effectively removed. Gmar et al. [57] showed the application of electrodialysis for the recovery of lithium from primary and secondary resources. Gally et al. [58] applied electrodialysis for the treatment of effluents from the Brazilian sewage treatment plant as a tertiary treatment. The results show 100% effectiveness to meet the quality standards of Brazilian legislation on the release of discharges. Lafi et al. [59] combined the ultrafiltration and electrodialysis process for the treatment of textile wastewater. The integrated ultrafiltration process reduced the COD, and due to this pretreatment the effluents do not clog the membrane during the electrodialysis process and this meets the reuse criteria.

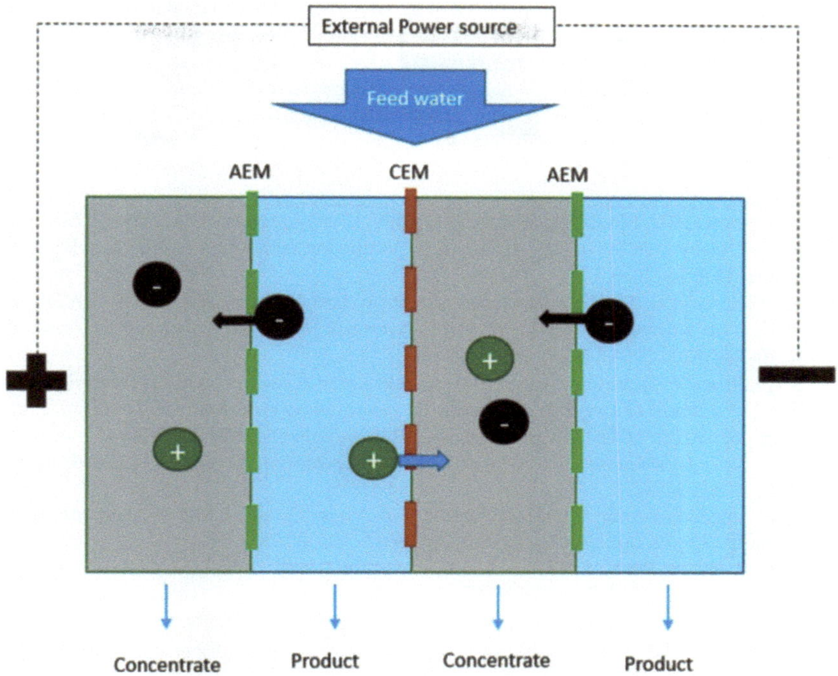

FIGURE 5.10 Schematic of electrodialysis process. From N. Mir, Y. Bicer, Integration of electrodialysis with renewable energy sources for sustainable freshwater production: a review, J. Environ. Manag. 289 (2021) 112496, *https://www.doi:10.1016/j.jenvman.2021.112496.*

Though electrodialysis is widely used, still some challenges and improvements are necessary for acceptance in the industry. Some challenges exist like increasing the membrane life, decreasing the cost of the membrane, improving the antifouling properties, and increasing the selectivity as well [60].

References

[1] World Health Organization, Progress on Household Drinking Water, Sanitation and Hygiene 2000-2020: Five Years into the SDGs, 2021.

[2] C. Sirtori, A. Zapata, I. Oller, W. Gernjak, A. Agüera, S. Malato, Decontamination industrial pharmaceutical wastewater by combining solar photo-Fenton and biological treatment, Water Res. 43 (3) (2009) 661–668.

[3] T. Wintgens, F. Salehi, R. Hochstrat, T. Melin, Emerging contaminants and treatment options in water recycling for indirect potable use, Water Sci. Technol. 57 (1) (2008) 99–107.

[4] A. Gulkowska, H.W. Leung, M.K. So, S. Taniyasu, N. Yamashita, L.W. Yeung, B.J. Richardson, A.P. Lei, J.P. Giesy, P.K. Lam, Removal of antibiotics from wastewater by sewage treatment facilities in Hong Kong and Shenzhen, China, Water Res. 42 (1–2) (2008) 395–403.

[5] S. Suárez, M. Carballa, F. Omil, J.M. Lema, How are pharmaceutical and personal care products (PPCPs) removed from urban wastewaters? Rev. Environ. Sci. Biotechnol. 7 (no. 2) (2008) 125–138.
[6] B.R. Bradley, G.T. Daigger, R. Rubin, G. Tchobanoglous, Evaluation of onsite wastewater treatment technologies using sustainable development criteria, Clean Technol. Environ. Policy 4 (2) (2002) 87–99.
[7] W. Viessman, M.J. Hammer, E.M. Perez, P.A. Chadik, Water Supply and Pollution Control, 1998.
[8] G.A. Engwa, P.U. Ferdinand, F.N. Nwalo, M.N. Unachukwu, Mechanism and health effects of heavy metal toxicity in humans. Poisoning in the Modern World-New Tricks for an Old Dog 10, 2019.
[9] S.K. Shukla, et al., Low-cost activated carbon production from organic waste and its utilization for wastewater treatment, Appl. Water Sci. 10 (2) (2020) 1–9, https://doi.org/10.1007/s13201-020-1145.
[10] S. Anandan, V. Kumar Ponnusamy, M. Ashokkumar, A review on hybrid techniques for the degradation of organic pollutants in the aqueous environment, Ultrason. Sonochem. 67 (2020) 105130, https://doi.org/10.1016/j.ultsonch.2020.105130.
[11] V.A. Vanoni, Sedimentation Engineering, Sedimentation Engineering, 2006. doi:10.1061/9780784408230.
[12] Q. Chang, Colloid and Interface Chemistry for Water Quality Control, Academic Press, 2016, pp. 23–35. doi:10.1016/B978-0-12-809315-3.00003-7.
[13] H.H. Steinour, Rate of sedimentation, Ind. Eng. Chem. Res. 36 (9) (2002) 840–849, https://doi.org/10.1021/IE50417A018.
[14] X. Yan, et al., A study on the mechanism of calcium ion in promoting the sedimentation of illite particles, J. Water Process. Eng. 42 (2021) 102153, https://doi.org/10.1016/J.JWPE.2021.102153.
[15] H. Anlauf, Recent developments in centrifuge technology, Sep. Purif. Technol. 58 (2) (2007) 242–246, https://doi.org/10.1016/J.SEPPUR.2007.05.012.
[16] S. Rahmat, F.M. Kusin, W.L. Wan Johari, Performance assessment of centrifuge dewatering unit using multivariate statistical approach: a case study of a centralized sludge treatment facility (CSTF) in Malacca, Malaysia, Desalination Water Treat. 57 (9) (2014) 3904–3915, https://doi.org/10.1080/19443994.2014.988650.
[17] C.F. Lin, Y. Shien, Sludge dewatering using centrifuge with thermal/polymer conditioning, Water Sci. Technol. 44 (10) (2001) 321–325, https://doi.org/10.2166/WST.2001.0652.
[18] J. Gregory, C.R. O'Melia, Fundamentals of flocculation, Crit. Rev. Environ. Sci. Technol. 19 (3) (1989) 185–230.
[19] F. Renault, B. Sancey, P.M. Badot, G. Crini, Chitosan for coagulation/flocculation processes—an eco-friendly approach, Eur. Polym. J. 45 (5) (2009) 1337–1348.
[20] C.S. Lee, J. Robinson, M.F. Chong, A review on application of flocculants in wastewater treatment, Process Saf. Environ. Protect. 92 (6) (2014) 489–508.
[21] A.K. Verma, R.R. Dash, P. Bhunia, A review on chemical coagulation/flocculation technologies for removal of colour from textile wastewaters, J. Environ. Manag. 93 (1) (2012) 154–168.
[22] T. Mustafa, T.K. Yagub, Dye and its removal from aqueous solution by adsorption: a review, Adv. Colloid Interface Sci. 209 (2014) 172–184, https://doi.org/10.1016/j.cis.2014.04.002.
[23] L. Guilherme, G.M. Dotto, Current scenario and challenges in adsorption for water treatment, J. Environ. Chem. Eng. 8 (2020) 103988, https://doi.org/10.1016/j.jece.2020.103988.
[24] V.K. Imran Ali, Advances in water treatment by adsorption technology, Nat. Protoc. 1 (2006) 2661–2667, https://doi.org/10.1038/nprot.2006.370.

[25] V.K. Gupta, Application of low-cost adsorbents for dye removal – a review, J. Environ. Manag. 90 (2009) 2313–2342, https://doi.org/10.1016/j.jenvman.2008.11.017.
[26] A. Bhatnagar, M. Sillanpää, Utilization of agro-industrial and municipal waste materials as potential adsorbents for water treatment—a review, Chem. Eng. J. 157 (2–3) (2010) 277–296.
[27] N.B. Singh, G. N., Water purification by using adsorbents: a review, Environ. Technol. Innovat. 11 (2018) 187–240, https://doi.org/10.1016/j.eti.2018.05.006.
[28] A.G. Niyousha Karbalaei Mohammad, Hydroxide modified activated alumina as an adsorbent for CO2 adsorption: experimental and modeling, Int. J. Greenh. Gas Control 88 (2019) 24–37, https://doi.org/10.1016/j.ijggc.2019.05.029.
[29] R.M. Bhargavi, Synthesis and use of alumina nanoparticles as an adsorbent for the removal of Zn(II) and CBG dye from wastewater, Int. J. Integrated Care 6 (2015) 31–41, https://doi.org/10.1007/s40090-014-0029-1.
[30] M.A. Rahman, Removal of methylene blue from waste water using activated carbon prepared from rice husk, Dhaka Univ. J. Sci. 60 (2012) 185–189, https://doi.org/10.3329/dujs.v60i2.11491.
[31] S.R. Sahar Abbaszadeh, Treatment of lead-contaminated water using activated carbon adsorbent from locally available papaya peel biowaste, J. Clean. Prod. 118 (2016) 210–222, https://doi.org/10.1016/j.jclepro.2016.01.054.
[32] K. Kadirvelu, Activated carbon from coconut coirpith as metal adsorbent: adsorption of Cd(II) from aqueous solution, Adv. Environ. Res. 7 (2) (2003) 471–478, https://doi.org/10.1016/S1093-0191(02)00018-7.
[33] N.R. Sudheer Kumar Shukla, Low-cost activated carbon production from organic waste and its utilization for wastewater treatment, Appl. Water Sci. 10 (2020) 62, https://doi.org/10.1007/s13201-020-1145-z.
[34] D. Saidulu, A. Majumder, A.K. Gupta, A systematic review of moving bed biofilm reactor, membrane bioreactor, and moving bed membrane bioreactor for wastewater treatment: comparison of research trends, removal mechanisms, and performance, J. Environ. Chem. Eng. 9 (5) (2021) 106112.
[35] S.D. Priyadharshini, P.S. Babu, S. Manikandan, R. Subbaiya, M. Govarthanan, N. Karmegam, Phycoremediation of wastewater for pollutant removal: a green approach to environmental protection and long-term remediation, Environ. Pollut. 290 (2021) 117989.
[36] T.J. Pfeiffer, P.S. Wills, Evaluation of three types of structured floating plastic media in moving bed biofilters for total ammonia nitrogen removal in a low salinity hatchery recirculating aquaculture system, Aquacult. Eng. 45 (2) (2011) 51–59.
[37] H. Odegaard, Innovations in wastewater treatment: the moving bed biofilm process, Water Sci. Technol. 53 (9) (2006) 17–33.
[38] X.J. Wang, S.Q. Xia, L. Chen, J.F. Zhao, N.J. Renault, J.M. Chovelon, Nutrients removal from municipal wastewater by chemical precipitation in a moving bed biofilm reactor, Process Biochem. 41 (2006) 824–828.
[39] M. Kermani, B. Bina, H. Movahedian, M.M. Amin, M. Nikaein, Application of moving bed biofilm process for biological organics and nutrients removal from municipal wastewater, Am. J. Environ. Sci. 4 (6) (2008) 675.
[40] L. Chu, J. Wang, Comparison of polyurethane foam and biodegradable polymer as carriers in moving bed biofilm reactor for treating wastewater with a low C/N ratio, Chemosphere 83 (1) (2011) 63–68.
[41] S. Zhang, Y. Wang, W. He, M. Wu, M. Xing, J. Yang, N. Gao, M. Pan, Impacts of temperature and nitrifying community on nitrification kinetics in a moving-bed biofilm reactor treating polluted raw water, Chem. Eng. J. 236 (2014) 242–250.
[42] J. Bundschuh, M. Kaczmarczyk, N. Ghaffour, B. Tomaszewska, State-of-the-art of renewable energy sources used in water desalination: present and future prospects, Desalination 508 (2021) 115035.

[43] V.K. Chauhan, S.K. Shukla, J.V. Tirkey, P.K.R. Singh, A comprehensive review of direct solar desalination techniques and its advancements, J. Clean. Prod. 284 (2021) 124719.
[44] F. Alnaimat, M. Ziauddin, B. Mathew, "A review of recent advances in humidification and dehumidification desalination technologies using solar energy," Desalination (2021) 114860.
[45] Y.-D. Chen, X. Duan, X. Zhou, R. Wang, N.-Q. Ren, S.-H. Ho, Advanced oxidation processes for water disinfection: features, mechanisms and prospects, Chem. Eng. J. 409 (2021) 128207.
[46] W.T. Vieira, M. Barbosa de Farias, M.P. Spaolonzi, M.G. Carlos da Silva, M.G.A. Vieira, Latest advanced oxidative processes applied for the removal of endocrine disruptors from aqueous media – a critical report, J. Environ. Chem. Eng. 9 (4) (2021) 105748.
[47] S. Anandan, V.K. Ponnusamy, M. Ashokkumar, A review on hybrid techniques for the degradation of organic pollutants in aqueous environment, Ultrason. Sonochem. 67 (2020) 2020, 105130.
[48] Y.J. Lim, K. Goh, M. Kurihara, R. Wang, Seawater desalination by reverse osmosis: current development and future challenges in membrane fabrication – a review, J. Membr. Sci. 629 (2021) 119292.
[49] H. Saleem, S.J. Zaidi, Nanoparticles in reverse osmosis membranes for desalination: a state of the art review, Desalination 471 (2020) 114171.
[50] A. Matin, T. Laoui, W. Falath, M. Farooque, Fouling control in reverse osmosis for water desalination & reuse: current practices & emerging environment-friendly technologies, Sci. Total Environ. 765 (2021) 142721.
[51] J. Kavitha, M. Rajalakshmi, A.R. Phani, M. Padaki, Pretreatment processes for seawater reverse osmosis desalination systems—a review, J. Water Proc. Eng. 32 (2019) 100926.
[52] S. Al-Amshawee, M.Y.B.M. Yunus, A.A.M. Azoddein, D.G. Hassell, I.H. Dakhil, H.A. Hasan, Electrodialysis desalination for water and wastewater: a review, Chem. Eng. J. 380 (2020) 122231.
[53] N. Mir, Y. Bicer, Integration of electrodialysis with renewable energy sources for sustainable freshwater production: a review, J. Environ. Manag. 289 (2021) 112496, https://doi.org/10.1016/j.jenvman.2021.112496.
[54] R. Mohammadi, W. Tang, M. Sillanpää, A systematic review and statistical analysis of nutrient recovery from municipal wastewater by electrodialysis, Desalination 498 (2021) 114626.
[55] K.S. Barros, T. Scarazzato, V. Pérez-Herranz, D.C.R. Espinosa, Treatment of cyanide-free wastewater from brass electrodeposition with edta by electrodialysis: evaluation of underlimiting and overlimiting operations, Membranes 10 (4) (2020) 69.
[56] L.M. Louidi Lauer Albornoz, Electrodialysis applied to the treatment of an university sewage for water recovery, J. Environ. Chem. Eng. 7 (2019) 102982, https://doi.org/10.1016/j.jece.2019.102982.
[57] S. Gmar, A. Chagnes, Recent advances on electrodialysis for the recovery of lithium from primary and secondary resources, Hydrometallurgy 189 (2019) 105124.
[58] T.B.-F.-H. Caline Rodrigues Gally, Electrodialysis for the tertiary treatment of municipal wastewater: efficiency of ion removal and ageing of ion exchange membranes, J. Environ. Chem. Eng. 6 (5) (2018) 5855–5869, https://doi.org/10.1016/j.jece.2018.07.052.
[59] L.G. Ridha Lafi, Treatment of textile wastewater by a hybrid ultrafiltration/electrodialysis process, Chem. Eng. Process. 132 (2018) 105–113, https://doi.org/10.1016/j.cep.2018.08.010.
[60] T.X. Chuanhui Huang, Application of electrodialysis to the production of organic acids: state-of-the-art and recent developments, J. Membr. Sci. 288 (2007) 1–12, https://doi.org/10.1016/j.memsci.2006.11.026.
[61] Y. Zhang, M. Sivakumar, S. Yang, K. Enever, M. Ramezanianpour, Application of solar energy in water treatment processes: a review, Desalination (2018) 116–145.

[62] C.V. Rekhate, J.K. Srivastava, Recent advances in ozone-based advanced oxidation processes for treatment of wastewater- A review, Chem. Eng. J. Adv. 3 (2020) 100031.
[63] L. Qiu, J. Liu, M. Li, S. Wang, W. Bai, D. Zhang, Assessment of heavy metal contamination in surface sediments from the nearshore zone, southern Jiangsu Province, China, Mar. Pollut. Bull. 133 (2018) 281–288, https://doi.org/10.1016/j.marpolbul.2018.05.055.

Further reading

[1] D.K. Mohamad Amran Mohd Salleh, Cationic and anionic dye adsorption by agricultural solid wastes: a comprehensive review, Desalination 280 (2011) 1–13, https://doi.org/10.1016/j.desal.2011.07.019.
[2] Q. Ma, Z. Xu, R. Wang, Distributed solar desalination by membrane distillation: current status and future perspectives, Water Res. 198 (2021) 117154.
[3] W. Lee, I. Kang, C. Lee, Factors affecting filtration characteristics in membrane-coupled moving bed biofilm reactor, Water Res. 40 (9) (2006) 1827–1835.
[4] B. Rusten, M. McCoy, R. Proctor, J.G. Siljudalen, The innovative moving bed biofilm reactor/solids contact reaeration process for secondary treatment of municipal wastewater, Water Environ. Res. 70 (5) (1998) 1083–1089.
[5] G. Proano-Pena, A.L. Carrano, D.M. Blersch, Analysis of very-high surface area 3D-printed media in a moving bed biofilm reactor for wastewater treatment, PLoS One 15 (8) (2020) e0238386.

CHAPTER

6

3D-Printed membrane for water treatment

A. Manmadhachary[1], *Roopa Tulasi*[2], *and Priyanka Chattoraj*[1]

[1] **Department of Mechatronics, ICFAI Tech, Faculty of Science and Technology, ICFAI Foundation for Higher Education, Hyderabad, India;**
[2] **Department of Mechanical Engineering, Sreenidhi Institute of Science and Technology, Hyderabad, India**

6.1 Introduction

Water is a fundamental requirement for survival of all the living beings. With the increase in population, the resources are likely decreased but the need for water is increasing rapidly. Mostly in developed countries water is supplied through filtered units but it fails to provide for the entire nation. Furthermore, water gets decontaminated due to storage or transportation or many other reasons [1]. Hence, filtration plays an important role in increasing quality of water thus separating solid particles from water. Tons of wastewater is produced with the increase in population across industrial, agricultural sectors. Since mankind and its survival are water-dependent, freshwater requirement has increased drastically. This has led to concentrated competition for freshwater among all the sectors, especially for agriculture as if farms lack water it affects irrigation and reduces the production of livestock. Wastewater generation is inevitable and found in integral parts of every sector [2]. It was observed that in oil and gas industry, every one barrel of crude oil handled generates about 10 barrels of wastewater [3].

All the wastewater are fresh water with contaminants; if these can be removed by conventional techniques fresh water resources are enhanced and good quality of water can be made available to all. This seems to be the most obvious way to deal with water scarcity. Many methods are used in wastewater treatment; desalination plays a vital role in wastewater

3D Printing Technology for Water Treatment Applications
https://doi.org/10.1016/B978-0-323-99861-1.00007-2

treatment [4]. With respect to the limitations in conventional manufacturing, 3D printing (3DP) presents the probable to ensure prototyping of these materials. It is referred to as Additive technology or Rapid prototyping, 3DP is an addition of layers to fabricate objects. This procedure has no spatial constraints, i.e., even intricate strategies can be fabricated and accumulated in one operation [5]. It shows a clear advantage compared to formative or subtractive manufacturing techniques. By applying a broad spectrum of materials this technology has emerged in wider range of applications including engineering and even nonengineering [6,7]. Some of the mostly used materials are polymers, ceramics, and metals. Since then, much interest has been dedicated for the water treatment technologies such as materials using 3D printing for separation of membranes, water purification process, etc. Besides, applications of 3DP extend to usage in various industries like aerospace, automobile, construction, biomechanics, etc. With enormous applications 3DP is also recently applied to water treatment such as desalination and wastewater purification. Three-dimensional printing has engrossed research on the design and fabrication of membrane modules. Because 3DP technology has high flexibility for manufacturing, which is limited in traditional methods [8,9].

Also, in recent scenario attention has been paid to 3D printing technologies for imprinted materials for surface molecularities and applied in the areas of surface imprinting in wastewater treatment technology such as extracting solids and separating from water [10]. However, advancements in fabrication techniques for water treatment materials like membranes and separators which are considered as complex structure can be done by technique called surface imprint 3D embedded printing [11]. Imprinting techniques have more advantages compared to 3DP mainly in the design of polymer applications. In addition, using photocatalytic materials improves surface properties of membrane [12,13].

This chapter is mainly aimed at reviewing critical applications like membrane separation and wastewater treatment using surface imprint technology and 3DP technology. Similarly, various applications in the area of manufacturing of materials used in water treatment and membrane technologies are assessed while concluding with limitations and outlook for the future with its scalability and applicability.

6.2 Water pollutants

The major cause of water pollution in today's scenario has many reasons like improper disposal of waste, often dumped into water bodies untreated. Mostly untreated disposal of sewage causes pollution and waterborne diseases like cholera and diarrhea. Polluted water can also destroy plants and organisms living in and around aquatic eco systems. This can also harm the people and animals who consume it. Basically, there are four types of water pollutants like pathogens, macroscopic pollutants, inorganic compounds, and organic materials [14].

6.2.1 Pathogens

Pathogens can be virus, bacteria, or protozoa. Generally, bacteria can be found in water. The actual problem is when they are increased above the safe levels then the water is said to be contaminated. Most common bacteria found in water are *E. coli* and coliform. The presence of *E. coli* indicates the water is contaminated. However, coliform is present in environment, if it exceeds certain level then it is considered as a dangerous pollutant.

6.2.2 Macroscopic pollutants

These pollutants are generally large, visible to everyone. Primary source of macroscopic pollutant is trash, especially plastic waste. It is thrown in water bodies in large amounts illegally which end up contaminating rivers and oceans, thus destroying aquatic ecosystem. Other types of pollutants include industrial waste, pieces of wood, nurdles, etc.; however, these can be controlled.

6.2.3 Inorganic compounds

Inorganic compounds include heavy metals like mercury, cadmium, arsenic, copper, zinc, chromium, and barium. Although these are harmless in small quantities they turn into pollutants when found concentrated in water. Major reasons can be leaching through water and industrial accidents. They can cause severe health hazards in humans and other living organisms, including death.

6.2.4 Organic materials

These materials contain carbon molecules in their makeup. The main type of volatile organic chemical detected is methyl tert-butyl ether (MTBE). It was formerly used as air cleaning additive but it is now a banned chemical. Water contaminated with this chemical takes a lot of time to be completely purified. Health effects include leukemia, lymphoma, and tumors.

Currently, several techniques are used for the separation of water pollutants from wastewater. As of now a greater number of researchers are making efforts to introduce wastewater treatment techniques like filtration, coagulation-flocculation, and biological treatment over the years. Among these one of the most widely used technology is membrane technique. It has grown drastically in the past couple of decades as it aids in wastewater treatment. Although membrane technique is not a new development, the variable nature of wastewater creates scope for new innovations and enhancements, concerning effectiveness, energy, and technical skill requirements. To reduce membrane fouling which is considered as major challenge in membrane process, constant efforts in modification of membrane modules and membrane elements are required. This chapter discusses the uses of membrane technology in water treatment; its pros and cons are considered and overall applications of membrane fouling and membrane modules are discussed.

6.3 Membranes for wastewater treatment

Fundamentally, a membrane acts as a blockade which splits two segments from each other by constraining actions of components. Membranes are classified into isotropic and anisotropic. Isotropic membranes are physical structures uniform in composition whereas anisotropic membranes are nonuniform in structure. Microscopic membranes are generally isotropic and porous, due to its permeable nature; in this case permeation fluxes are relatively high when compared to nonporous membranes [15]. Alternatively, anisotropic membranes are nonuniform in nature and made up of varying layers with different compositions. These membranes are generally used in Reverse Osmosis (RO) process because it has a very thin layer reinforced with a thick and very penetrable layer [16–19].

Generally, membranes are divided into organic and inorganic. Synthetic polymers are used to make organic membranes. Frequently, these are pressure driven and used for water separation process like filtration, reverse osmosis, ultrafiltration, and nanofiltration. The materials used for these processes derived from synthetic organic polymers are polyethylene (PE), poly tetra fluor ethylene (PTFE), polypropylene, etc. The inorganic membranes are derived from materials like metals, ceramics, or silica [20].

Membrane fouling is accumulation of solids or microbes on the surface of the membranes, or even in the pores of the membranes, resulting in reduced permeability thereby reducing the efficiency of the membrane. Higher the fouling, the more pressure required. Overall membrane performance can be affected by membrane fouling. High energy consumption and reduction in membrane performance are the results of fouling [21].

Pretreatment is given to wastewater before the membrane separation process. It has various advantages like it contributes to energy utilization and also to make membrane more efficient; it maintains chemical and physical properties.

6.4 Wastewater treatment processes

6.4.1 Pressure-driven processes

Pressure-driven processes are the most widely used processes in wastewater treatment from pretreatment to post-treatment. Generally, hydraulic pressure is used to achieve separation. There are four different types involved in pressure-driven processes; they are microfiltration, ultrafiltration, nanofiltration, and reverse osmosis (RO) [22].

Among these pressure-driven processes, RO is used widely for its effectiveness in separating minor particles like bacteria and univalent ions like sodium and chloride up to 99.5%. It has been the lead process in treating wastewater and desalination of sea water from years on end [14–17].

Pressure-driven processes are undoubtedly innovative techniques to treat wastewater, but challenges still remain. With the requirements in energy due to the pressure, these can be removed by other process like forward osmosis process.

6.4.2 Forward osmosis

In forward osmosis (FO) process water molecules are moved from one solution to another through a partially porous membrane. Osmotic pressure is used as driving force. A solution of relatively high concentrated feed solution is used to induce net flow of water through the membrane into draw solution. This effectively separates feed water from its solutes [23] (Fig. 6.1).

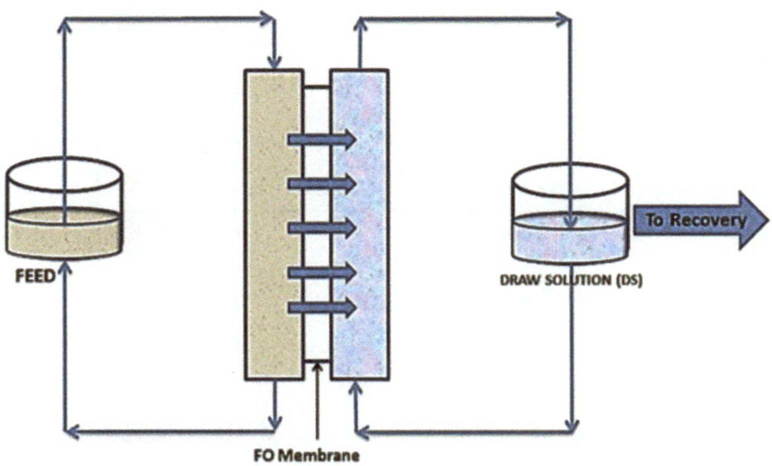

FIGURE 6.1 Schematic diagram of forward osmosis.

Forward osmosis has numerous advantages like there is no need of external pressure, hence, energy consumption reduces. Water cleaning and fouling are made easy due to osmotic pressure. Selecting draw solution makes the process flexible for customization of products for fresh water recovery. Furthermore, the reuse and regeneration of draw solution is advantageous for cost saving. Considering all the points FO can treat highly concentrated products henceforth opting the choice of forward osmosis as an improved one.

Apart from these processes for wastewater treatment in membrane technology many other processes are also used depending upon its nature and applications. Since this chapter is more involved into 3DP techniques for membrane technology and its modules, membrane modules and selection are discussed in further sections.

6.5 Membrane modules and selection

For large-scale membrane processes, large membrane areas are required; for industrial or commercial use these are packaged economically called as modules. Basically, these are four types of membrane modules like plate and frame, spiral, tubular, and hollow fiber [14–17]. The properties of these modules are shown in Table 6.1.

Membrane technology is gradually becoming the most used technology for wastewater treatment. However, there is always room for improvement and new technologies are being developed to overcome the drawbacks involved in membrane technology like draw solute recovery process to make cheaper, suitable membranes for large scale applications.

To overcome these drawbacks one of the latest innovations in manufacturing sector has introduced additive manufacturing techniques known as 3D printing technology for manufacturing of membranes with sustainable technology. This technology has already proved its efficiency

TABLE 6.1 Properties of various membrane modules.

Property	Plate and frame	Tubular	Spiral wound	Hollow fiber
Packing density (m^2/m^3)	148–492	20–374	492–1247	492–4924
Potential for fouling	Moderate	Low	High	Very high
Ease of cleaning	Good	Excellent	Poor	Poor
Relative manufacturing cost	High	High	Moderate	Low

in various sectors like automobile, aerospace, medical, and various engineering applications. Now it is now here less to show its efficiency in wastewater treatment process like fabrication of modules, spacers, membranes to obtain fresh water [24].

6.6 3D printing technology (additive manufacturing)

3DP or additive manufacturing (AM) is growing predominantly in several areas such as aerospace, automobile, manufacturing, and even healthcare. AM is also called as rapid prototyping, layered manufacturing, digital manufacturing, etc. Generally, additive indicates adding layer by layer in fabrication using digital data [25].

In 3DP process a 3D digital data of desired design is prepared by Computer Aided Design software and is converted into files like .STL or.OBJ, which are compatible for 3D printers. This data is then processed using slicing software which slices the model into several layers. Numerous parameters like printing speed, layer height, infill density, etc., are used to optimize the design process. The process of slicing converts .STL file into certain G codes which are accepted by certain 3D printers. This file is used to fabricate a part. Mostly, post-processing is needed to take away the support structures from the finished part [26]. These 3DP process steps are shown in Fig. 6.2.

There are different types of 3D printers used across the industries depending on the type of materials, type of methods, type of applications, etc. The following are major plastic-based 3D printers

- Material extrusion: Fused Deposition Modeling (FDM), Fused Filament fabrication (FFF), paste extrusion, etc.

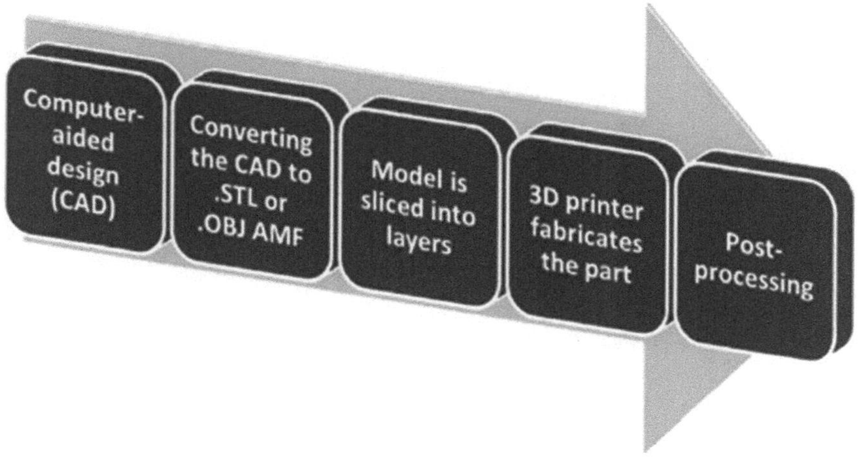

FIGURE 6.2 3DP process steps.

- Powder bed fusion: Selective Laser Sintering (SLS), Selective Laser Melting (SLM), Direct Metal Laser Sintering (DMLS), etc.
- Vat polymerization: Stereolithography (SLA), Digital light Processing (DLP), Continuous Liquid Interface Production (CLIP)

The most frequently used 3D printer is based on the principle called extrusion also known as FDM or FFF where thermoplastic materials are used as printing material in membranes. Here the printer melts the thermoplastic material according to desired sliced data to print a 3D object. The melt material is set layer on layer up until the desired entity is formed. Materials that are commonly used for FDM are polylactic acid (PLA), acrylonitrile butadiene styrene (ABS), polycarbonates (PC), poly ether ether ketone (PEEK), and many other polymers. Here, material gets into molten form when heated and solidifies after printing. This technique is flexible to operate and user friendly and less expensive and available in different sizes, e.g., desktop printer is also available in this method of printing [27].

Another common type of 3DP technology is Selective Laser Sintering (SLS) where powder is used as primary material to print the part. In this technology almost all parts like metals, ceramics, polymers, etc., can be printed without any use of support structures. The mechanisms of this process include a printing bed, build chamber, powder reserve chamber, and laser beam, which is used as a source for printing.

Stereolithography is another type of 3DP technique where a part is created by photosensitive polymer resins using an UV laser beam. High resolution is obtained using this technique. Post-processing may or may not be required depending on the complexity of the model. Compared to SLA, DLP is considered a faster process, as it uses a different type of laser called arc lamp with micromirrors. The light is applied to complete surface of thermoset photo polymer resin in single projection; this way it has flexibility to print object faster.

3DP plays a vital role in water treatment applications. It has potential to print millions of nanofibers and compress into a membrane. The role of 3DP helped a lot of researchers in membrane engineering, as it is difficult to obtain through conventional techniques such as electroblowing, electro spinning, etc. Since 3DP has potential to print microstructure membranes and also macrostructures, the efficiency of the final object can be determined by various factors. Some of them are;

- High priority must be given to commonly used materials for manufacture of membranes.
- High resolution is always an important factor.
- Precision and greater accuracy helps for error-free products.
- The membrane time for production should be maintained as low as possible.

The 3DP technique is considered as step-by-step approach; mostly a product can be finished in one go. This can reduce the dependency on other conventional or manufacturing techniques like welding and then coating using deposition methods. Three-dimensional printing techniques have several advantages over traditional methods [28].

6.7 3D printed materials for desalination and membrane separation

Three-dimensional printing technology has provided remarkable advancements in water treatment process like membrane design, fabrication of membranes, oil-water separation techniques, and materials for treatment of wastewater. Many researchers are manufacturing of parts for wastewater treatment because of its flexibility and ability to print complex structures [29,30]. List of various materials and methods are listed in Table 6.2.

6.8 Membrane separation

Channel feed spacers are useful for continuous flow and recirculation and also for fluid mixing. For many membrane separation applications these feed spacers are used especially in spiral wound membranes (SWM) [31]. These are made in different designs to enhance the area of turbulence, supporting in mechanical aspects for fluids to traverse, most importantly to reduce fouling, also to prevent mutilation of the layers in membranes [32]. The representation of how spacers and membranes are placed is shown in Fig. 6.3.

Design of virtuous spacer is important because dead zones can create deposition of particles. Hence it leads to fouling which can reduce the rate of mass transfer. Different spacer designs are manufactured and tested with various characteristics like pressure drop, effect on its mass transfer, and also to reduce dead zones [33]. However, there are certain difficulties in manufacturing of these spacers using conventional manufacturing techniques. Some of them are molding, vacuum forming, heat extrusion, etc. Hence, the potential of 3DP is identified to fabricate spacers like it can fabricate any simple and complex shapes.

Based on the literature it is observed that feed spacer materials are the most researched area for 3DP applications, especially for wastewater treatment and desalination. This is because of print resolution in present commercial 3D printers where intricate parts can be made-up. Utmost the spacers which are printed using 3DP show improved performance in aspects of fouling, performance of flux, improving hydrodynamic flow compared to conventionally made spacers. One of the drawbacks using

TABLE 6.2 3D Printing applications in water treatment.

Application in water treatment	3D printed parts	Material used	Technology used
Membrane separation			
Filtration	Spacers	Polypropylene	SLS
RO UF	Spacers	Polyamide-12	SLS
Microfiltration	Vibrating spacer	Polyamide	SLS
Ultrafiltration	Spacers	Polypropylene, polylactic acid (PLA)	FDM, POLYJET
Filtration	Spacers	Acrylonitrile butadiene styrene (ABS)	SLS
Membranes			
Filtration	Ceramic membranes	Clay powder, maltodextrin powder	Inkjet
Oil water separation	Membrane	Polysulfide, polyamide	SLS
Wastewater treatment			
MBBR	Biofilter carrier	Liquid resin	Polyjet
Solar steam generation			
Desalination	Solar evaporator	Graphene oxide and carbon nanotubes	FDM
Degradation of dyes			
Methylene blue removal	Metal organic frameworks composites	ABS layered with Cu-BTC	FDM
Oil water separation treatment			
Water oil separation	Oil skimmer mesh	E-glass with carbon nanotubes	SLA
Water oil separation	Ceramic mesh	Alumina	DLP

From Table 6.2, it is observed that 3DP has a tremendous role in wastewater treatment. In this chapter membrane and membrane separators are discussed in detail in further sections.

3D printers is poor surface finish of some materials. However, if this can be overcome by using more materials then 3DP in desalination and wastewater treatment will gain more attraction [34].

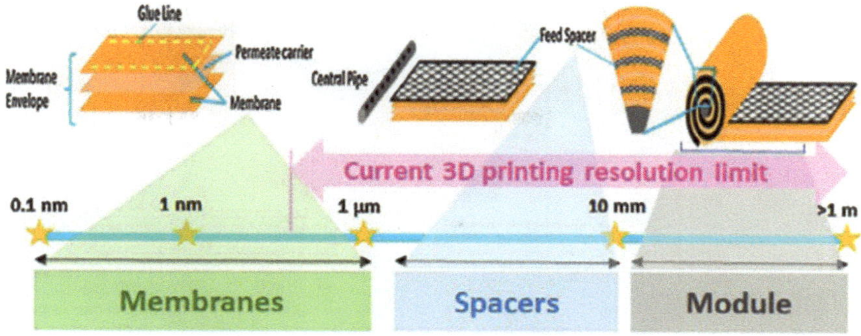

FIGURE 6.3 Spiral wound membrane including spacers and modules.

6.9 Membranes for filtration

Membrane technology is used rapidly due to its cost-effectiveness, high efficiency, etc., compared to other traditional methods. Most of them are polymers but ceramics can also be used for membranes. Traditional methods for polymers are phase inversion, extrusion, spinning, etc. Electrospinning technique is also found useful after many studies for wastewater treatment applications and desalination [35–37]. Then the capability of 3DP came into existence, but due to less materials and resolution limits this technology is not widely used directly for polymer materials. Most of the 3D printers used nowadays are not efficient to print in below micron resolution wherein the membrane pores are present at submicron levels. Furthermore to this, approximately membranes should have certain characteristics like wettability which are hydrophilic, hydrophobic, and good mechanical properties.

Research has been done to coat polyamide membrane substrate which is 3D printed in (SLS) with super hydrophobic structure. Basically, two steps were involved in fabricating this membrane. First was synthesis of the base material, second step was coating of unique pore system called Zeolitic imidazolate framework (ZiF) on the PA substrate. On using this membrane efficiency about 98% was achieved in oil water separation (Fig. 6.4) [38].

However, printing complete membrane using one appropriate 3DP technology is still in research but one application is oil water separation membrane was practically made possible by researchers. In this the inkjet printer is used to print membrane using PDMS nano silica ink which is porous and super hydrophobic. Here ink is prepared in a container with a

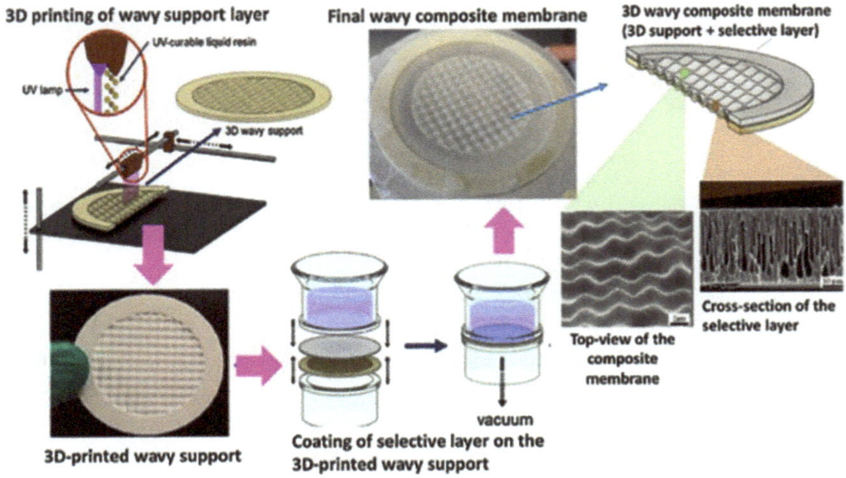

FIGURE 6.4 3D printed composite membrane.

stirrer containing silica particles with PDMS. This ink is printed on a glass substrate which is heated, using a screw pump, nozzle, and 3-axis movement-controlled platform. After curing, peel of the printed layer out of glass substrate to get a porous membrane. This method has a higher efficiency unlike traditional methods (Fig. 6.5) [27].

Furthermore, research is going on using 3DP in many other wastewater membranes using polymers. Nowadays 3DP is gaining attention in fabricating support structures for nanofibrous membranes. Using SLS techniques research has been done to fabricate support structures made of PET for membranes like PMMA, PDMS, etc. One of the researchers

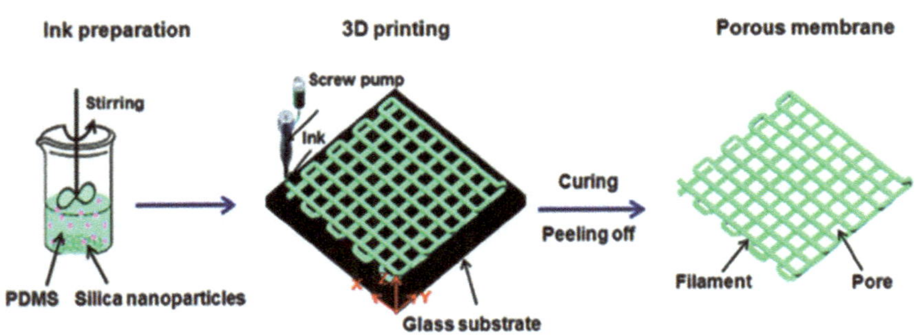

FIGURE 6.5 Membrane prepared by inkjet printer.

inspired from lotus leaf to fabricate porous hydrophobic PLA membrane for oil water separation using FDM. Its efficiency was observed around 96% with minimum pore size [31,38].

There is significant increase in research to fabricate the parts of membrane technology by 3DP. But, the actual resolutions of practical 3D printers are not ample to print pore sizes of membranes and required tolerances. However, 3D printers like two photon polymerizations are able to print submicronic levels in resolution, still restricted to consistent printing and surface quality, e.g., the obtained model is not accurate as premediated. Likewise, several membranes require very fine and small pores for ultrafiltration, nanofiltration, and reverse osmosis processes [39,40]. This process with these limitations perhaps needs further research in membrane designs using composites instead not to print straight on the membrane. Further important aspect to be considered here is to prevent delamination it is essential to have good compatibility between active layer and support layer.

6.10 Capsules or biocarriers for wastewater

Wastewater is the main source for microorganisms, pharmaceuticals, and dyes which are difficult to remove from water using conventional techniques. Post-treatment is generally used to treat these sewages. One of the traditional techniques used to treat or breakdown waste is chemical oxidation. This technique uses ferrite VI to breakdown organic and inorganic compounds. It has shown the maximum efficiency. But it is highly expensive to produce and it will not be stable in moist atmosphere. By easy encapsulation, it can be made more stable and cost-effective, and it can be made resilient; its release can be controlled. This controlled-release strategy is found efficient in removing dissolved contaminants from wastewater [41].

The researchers applied 3DP technique to print capsules used for wastewater treatments. These capsules are made of poly vinyl alcohol (PVA) to encapsulate ferrite VI and compared with traditionally made capsules [41]. The results show that 3D printed capsules are more efficient in long term like storing for almost a month without loss in its efficiency. This encapsulation technique using 3DP is useful in emergency to store or apply ferrite. Multistep process is involved here where printing of capsule and fill-up ferrite inside the capsule. This is one of the major limitations. A one-step unique strategy like straight away printing a capsule using ferrate maybe achieved in near future. Multimaterial printers make possible to achieve this limitation [42] (Fig. 6.6).

FIGURE 6.6 Capsules made for posttreatment process (A) commercially made gelatine capsules (B) 3D printed PVA capsules.

6.11 Substrates for oil-water separation

One of the main environmental concerns in today's scenario is oil pollution and spillage. The main reason for pollution is oil exploitation, oil transportation, and extraction. Oil-water separation technology has acquired importance for oil clean-up during spillage. Among all methods and materials used for oil cleanup one of the efficient methods found is porous structure membrane. It has the ability to recycle. The porous material allows oil to pass but hinders water at its surface. This is due to its properties like super hydrophobic also called as wettability and apex surface to unit volume ratio. But traditional manufacturing methods are slow moving and difficult to fabricate intricate shapes [43–45]. This is where 3DP applications gained a special attraction due to their ability to print complex shapes and fabricate porous membranes with admissible properties for oil water separation technique. It is reported that both polymeric and ceramic materials are used to fabricate porous membrane structures.

The main challenge is coating at a nano level on the super hydrophobic structure for oil-water separation. By using 3DP technology studies found that polydimethylsiloxane (PDMS) is an ink containing hydrophobic nano silica. It is deposited on a mesh-like structure by 3DP. The existence of nano silica gives virtuous printability, good mechanical strength on the coated material. Thus, water-oil separation was achieved at a rate of 99.6% with minimum pore size. Mostly polymer-based materials are used for oil-water separation materials or substrates. But, one of the limitations of using polymers is when they are exposed to inharmonious conditions they may degrade. To overcome these limitations ceramic-based materials are introduced. This inspired researches to develop ceramic based like alumina materials for oil water separation. It is observed the efficiency is >99% even in high flow rate. Interestingly, it showed high efficiency when exposed to harsh conditions. Their material showed better durability compared to metals and polymers. The final 3D printed part can be optimized using heat treatments (Fig. 6.7) [46,47].

One of the important aspects to be considered when using 3DP is the mechanical integrity of the absorption material—the interfacial bond between 3D printed materials and coated materials. These nanoparticle inclusions are considered as the sorbent or substrate will be exposed to environmental challenges and also various loadings. However, method of printing and orientation of the build chamber are important factors when mechanical properties are considered [48]. The future may be printing sorbent materials through advanced technologies like 4D printing wherein multifunctionalities are added.

6.12 Dye degeneration in water treatment

Organic dyes are mostly carbon-based. These carbon-based absorbents or substrates are quite difficult for removal of dyes. However, the process of recycling and separation is actually cumbersome. This is because it lacks flexibility due to the materials which are used in powder form. Flexible and porous structures where organic dyes are directly adsorbed with high efficiency and malleable in recycling can be fabricated using 3DP technique. Metal organic frameworks (MOFs) are one of the promising materials which have high surface areas, open channels, porous crystalline structures. These inherent properties are used to remove carbon-based dyes. Toward the improvement in the elasticity of MOF using 3DP technique for the manufacture of substrate using materials like acrylonitrile butadiene styrene (ABS) deposited with porous Cu-benzene tricarbonyl acid adsorbents for removal of dye known as Methylene blue (MB). In this preparation coating of Cu-BTA on to a 3D printed ABS surface is done. This composite enhances its mechanical and chemical properties like wettability which helps to adsorb the metals (Fig. 6.8) [49,50].

FIGURE 6.7 Graphical representation of oil-water separator.

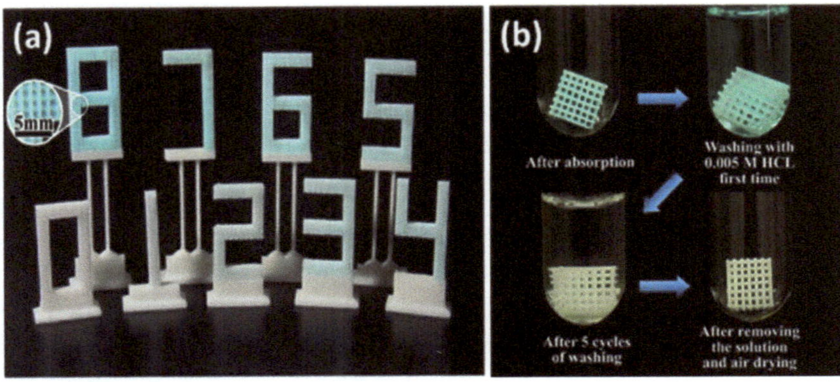

FIGURE 6.8 (A) Synthesized Cu-BTC/ABS composites (B) recycling process of ABS skeleton.

Substrates used for dye deprivation require high surface area to improve their efficiency. Practically, mechanical stability is an important factor for cyclic use of adsorbents. Many a times adsorbent materials are changed by modifiers like clay, CNT, etc., by coating or mixing. The main challenge here is it leads to poor adhesion of coated layers to the substrates which results in delamination. Challenges still exist like to follow precise digital model at the maximum resolution. Efforts are required in future to enhance the bond or adhesion between fillers and polymer matrix [51].

6.13 Use of graphene aerogels in water filtration

Three-dimensional printing has no barriers in treating wastewater, especially in purification and filtration. Research has been done in developing an innovative material like graphene in the use of filtration process. Graphene aerogels play a prominent role in large-scale purifying plants. It acts as a membrane for inorganic components, dyes, metals, etc., with maximum efficiency from drinking water. Furthermore, these graphene nanosheets are different from others because of their unique properties like they can be recycled and used multiple times without any residues left after use, they can be 3D printed for large-scale applications [51].

The main aim of this membrane is to eliminate impurities from water without leaving any chemical residues. The kind of aerogels prepared when put into water it retains its shape and structure in this way; this can be used in miscellaneous water treatment applications. The potential of graphene in water purification has high surface area and it has ability to

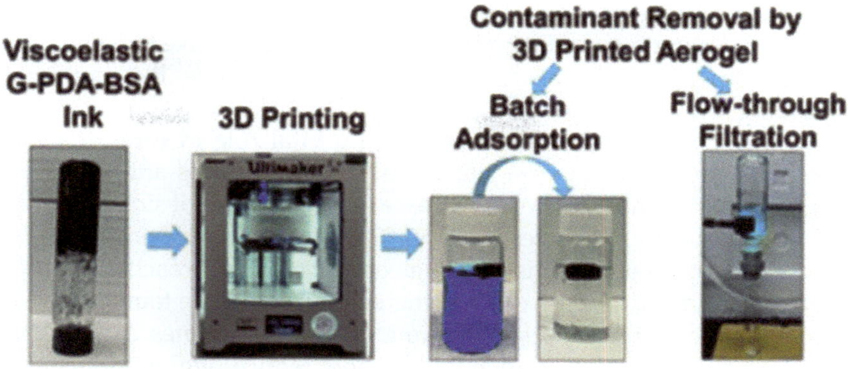

FIGURE 6.9 Process of graphene aerogel.

exert hydrophobic and electrostatic behavior due to which it has the ability to absorb contaminants from waste water. Compared to conventional techniques 3DP has already proved its efficiency in thermal and chemical properties; so it could be used in mass production (Fig. 6.9).

Use of 3D graphene porous structures is widely used in water contamination process. It has great thermal and mechanical properties so it gains more attraction toward absorption of contaminants from water. Basically, this procedure involves addition of two biopolymers to graphene ink to achieve high ability to print, namely PDA and BSA. These polymers help to remove contaminants. So together they are called as G-PDA-BSA aerogel. It is printed using Ultimaker desktop 3D printer. The printed graphene porous microstructure showed high performance in adsorption and effectively absorbs chemicals, organic dyes, inorganic solvents, etc. It also reduces methyl blue by 20%. It can be reused up to 15 cycles. This electrostatic behavior helps to remove heavy metals from water and one of the main advantages involved here is it can be made from small-scale to large-scale plants like scalability. After genuine research in this area, it can be stated that it is efficient to use graphene-based microporous structures in water treatment applications using 3DP [28].

6.14 Use of ceramics in water treatment

Basically, 3DP technique is divided into two types namely rapid manufacturing or direct methods and indirect methods. In rapid manufacturing the part is printed based on digital data and obtained product can be directly used as final object wherein there is no need of any post-processing. Whereas indirect method involves little or more post-processing depending upon the applications like removal of support structures, sintering, heat treatments if necessary, curing, etc.

These postprocessing procedures can reduce accuracy of the final part. There are some adverse effects especially in procedures like SLS and SLA. Usually, ceramic materials need postprocessing steps like sintering, debinding, etc. [52].

Generally, sintering temperature plays a vital role in ceramic membrane fabrication. It can straightaway affect the surfaces and structural parameters of membranes. With increase in temperature of sintering there will be increase in pore sizes but eventually it decreases porosity of the membrane. To overcome this challenge one or more approaches are used like to improve the kinetics of sintering and to reduce the temperature of sintering process for especially ceramic-based membranes. To enhance the performance, selection of powders and reduce agglomeration can probably improve the efficiency and utilizing additives is also an approach. So as far as ceramic membranes using 3DP are concerned, these approaches can maintain sintering temperatures for fabrication [53].

Another important aspect is debinding; it is considered as one of the main steps in production especially in powder metallurgy and ceramic components. Generally, it is a process of removing additives prior to fabrication techniques. Binder components like insoluble and soluble can be removed by thermal debinding process. This process can help to improve the quality of the membrane [54].

Typically, one more postprocessing technique in 3DP of ceramic membranes is infiltration. This method is used to block the pores to get greater mechanical strength to the obtained product, i.e., membranes. It is a part of post-processing; however, this can be done using any metals or wax so that the parts which are delicate can gain strength and made stable mechanically [55,56].

Basically, before selecting suitable techniques for membrane fabrication various parameters are considered like which method is suitable for membrane fabrication, fabricating cost, production setup, etc. After overall review the key parameters are evaluated, various 3DP techniques are used to produce ceramic membranes for processes like reverse osmosis, ultrafiltration, nanofiltration, etc. [42,57].

6.15 Challenges of 3D printing

6.15.1 Resolution/accuracy

Layer height is one of the disadvantages in membrane fabrication when printed directly, where layer height and pore size are in sub micrometer level. The two-photon polymerization technique would help in addressing this issue as it can print nanometer level [25].

6.15.2 Limited materials

More number of 3D printed materials are needed like polyvinylidene fluoride (PVDF), polyimide, polyetherimides, polytetrafluoroethylene (PTFE), etc., which are used in conventional membrane fabrication. The material used and type of technique result in efficiency of the final product. Mechanical strength is also an important factor where parts printed on 3D printer should withstand loads and pressures under challenging environments. This is especially necessary in wastewater treatments where it should deal with high pH levels. Some of the polymers like photopolymers undergo swelling when reacted with wastewater due to its high hydrophilic behavior.

6.15.3 Poor scalability/slow process

Although 3DP prints complicated structures, the time taken to print a large structure is relatively high. New 3DP technologies like CLIP can address this issue. SLS-AM produces rough surfaces. This can be sometimes advantageous and sometimes a disadvantage to membrane design and especially affects its fouling or antifouling properties.

6.15.4 Budget

3DP techniques are moderately quite affluent compared to various traditional methods. It is necessary to vividly cut the cost of the materials and process.

The prospective of additive printing in the near future appears to be unrestricted if completely the encounters are imparted. There will be more innovative technologies in near future due to 3D printers which are precise, inexpensive, smart materials, and to reduce preprocessing and post-processing high quality automated controlled printers are required.

6.16 The outlook for future

6.16.1 Combination of conventional methods with additive methods

Three-dimensional printing's future maybe more of a multiintegrated process rather than stand-alone process. This is predominantly important in wastewater treatment for membrane fabrication and desalination, as 3DP tenacity is not yet capable to directly pattern the substrates; thus an amalgamation of active layer using traditional methods and support layers using 3D Printing techniques is preferable.

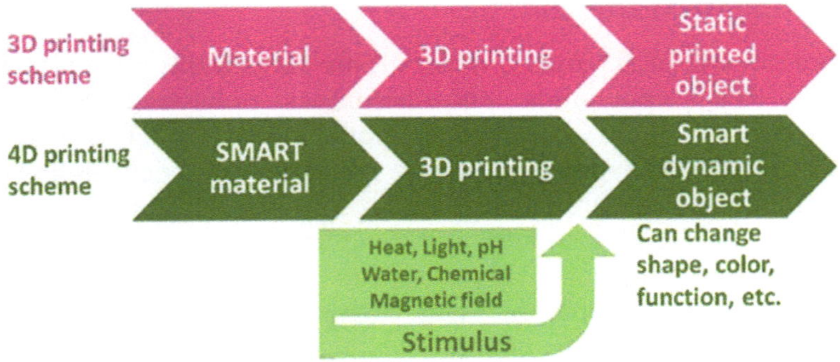

FIGURE 6.10 Difference between 3D printing and 4D printing.

6.16.2 3D printing + electrospinning/surface coatings

One of the innovative ways of fabricating a composite is by combining 3DP with surface coatings. This technique is mostly used in biomedical field rather than water treatment or desalination techniques. Three-dimensional printing with electrospinning is extensively used in multi-functional layered scaffolds used in bone reconstruction and nasal cartilages.

6.16.3 4D printing

Another exciting and innovative research in future is 4D printing. It is used to fabricate parts for wastewater treatment applications. This is upgradation of 3DP where addition of the element of time like property with respect to time would be possible. Thus, it delivers advantages like reduction in volume for storage, shape transformations, etc. These applications can be used in medicine, aerospace, satellites, etc. (Fig. 6.10).

6.17 Conclusions

Three-dimensional Printing technology has potential to fabricate flexible and intricate structures. These are used in various prototyping and fabrications including wastewater treatment technologies. It is versatile, fast, and efficient, which can fabricate any shape and geometry enabling a new standard in manufacturing industry. This review presented an overview of various exciting technologies in water treatment including the use of feed spacers, membranes, oil-water separators. It is observed that in most water treatment process with macro level materials the use of 3DP is suitable, as it has high degree of freedom. However, challenges are faced even in this technique especially on direct printing membranes. Key areas that need to be investigated are scalability, availability of materials, cost, printing speed, etc. Making the 3D printed material to be responsive to stimuli via 4D printing will open new prospects of research. Overall, the exciting field of manufacturing, i.e., 3DP is a new paradigm in new material design and fabrication that has wide applications in water treatment. There is already a radical advancement in 3DP applications; however, cost, process, materials, environmental challenges, reducing post-processing are to be addressed to fully use its unlimited potential.

References

[1] M.P. Barry, E.D. Kristen, H.R. Irwin, Water. Hydration and health, Nutr. Rev. 68 (8) (2010) 439–458.

[2] E.O. Ezugbe, S. Rathilal, Membrane technologies in wastewater treatment: a review, Membranes 10 (2020) 89.

[3] S.E. Nascimben, Z. Laszlo, C. Hodur, G. Arthanareeswaran, G. Verrb, Photocatalytic membrane filtration and its advantages over conventional approaches in the treatment of oily wastewater: a review, Asia Pac. J. Chem. Eng. 15 (5) (2020) 1–29.

[4] P.A. Sosa-Fernandez, J.W. Post, H. Bruning, F.A.M. Leermakers, H.H.M. Rijnaarts, Electro dialysis-based desalination and reuse of sea and brackish polymer-flooding produced water, Desalination 447 (2018) 120–132.

[5] ASTM F2792-12a, Standard Terminology for Additive Manufacturing Technologies, ASTM International, West Con-shohocken, PA, 2012.

[6] A. Manmadhachary, M.M. Aditya, R. Haranadha, Manufacturing of customized implants for orbital fractures using 3D printing, Bioprinting 21 (2021) e00118.

[7] A. Manmadhachary, K.M. Santosh, R.Y. Kumar, Design & manufacturing of spiral intake manifold to improve volumetric efficiency of injection diesel engine by AM process, Mater. Today Proc. 4 (2017) 1084–1090.

[8] Z.X. Low, Y.T. Chua, B.M. Ray, D. Mattia, I.S. Metcalfe, D.A. Patterson, Perspective on 3D printing of separation membranes and comparison to related unconventional fabrication techniques, J. Membr. Sci. 523 (2017) 596–613.

[9] N. Yanar, P. Kallem, M. Son, H. Park, S. Kang, H. Choi, A New era of water treatment technologies: 3D printing for membranes, J. Ind. Eng. Chem. 91 (2020) 1–14.

[10] N. Li, H. Yang, Construction of natural polymeric imprinted materials and their applications in water treatment: a review, J. Hazard Mater. 403 (2021) 123643.

[11] F. Cui, Z. Zhou, H.S. Zhou, Molecularly imprinted polymers and surface imprinted polymers based electrochemical biosensor for infectious diseases, Sensors 20 (4) (2020) 996.

[12] H.A. Balogun, R. Sulaiman, S.S. Marzouk, A. Giwa, S.W. Hasan, 3D printing and surface imprinting technologies for water treatment: a review, J. Water Proc. Eng. 3 (2019) 100786.

[13] X. Zhou, C.J. Liu, Three-dimensional printing for catalytic applications: current status and perspectives, Adv. Funct. Mater. 27 (30) (2017) 1–13.

[14] K.L. Jepsen, M.V. Bram, S. Pedersen, Z. Yang, Membrane fouling for produced water treatment: a review study from a process control perspective, Water 10 (7) (2018) 847.

[15] N.N. Bui, M.L. Lind, E.M.V. Hoek, J.R. McCutcheon, Electro spun nanofiber supported thin film composite membranes for engineered osmosis, J. Membr. Sci. 385–386 (1) (2011) 10–19.

[16] K. Guo-dong, C. Yi-ming, Development of antifouling reverse osmosis membranes for water treatment: a review, Water Res. 46 (3) (2012) 584–600.

[17] K.L. Tu, A.R. Chivas, L.D. Nghiem, Chemical cleaning effects on properties and separation efficiency of an RO membrane, Membr. Water Treat. 6 (2) (2015) 141–160.
[18] M. Duan, Z. Wang, J. Xu, J. Wang, S. Wang, Influence of hex methyl phosphor amide on polyamide composite reverse osmosis membrane performance, Separ. Purif. Technol. 75 (2) (2010) 145–155.
[19] S. Jiang, Y. Li, B.P. Ladewig, A review of reverse osmosis membrane fouling and control strategies, Sci. Total Environ. 595 (2017) 567–583.
[20] M. Yao, Y.C. Woo, L.D. Tijing, W.G. Shim, J.S. Choi, S.H. Kim, H.K. Shon, Effect of heat-press conditions on electro spun membranes for desalination by direct contact membrane distillation, Desalination 378 (2016) 80–91.
[21] S.A. El-Behary, A.M. El-Aassar, M. Aboelfadl, N. Abdel-Atey, A.T. Kandil, Surface modification of TFC membranes using PVA and nanomaterials for enhancing fouling resistance, Desalination Water Treat. 193 (2020) 106–116.
[22] M. Padaki, A.M. Isloor, K.K. Nagaraja, H.S. Nagaraja, M. Pattabi, Conversion of microfiltration membrane into nanofiltration membrane by vapour phase deposition of aluminum for desalination application, Desalination 274 (2011) 177–181.
[23] K. Jasmina, B. Subhankar, B. Malini, C. Helix-Nielsen, P. Irena, Forward osmosis in wastewater treatment processes, Acta Chim. Slov. 64 (1) (2017) 83–94.
[24] R. Godina, I. Ribeiro, F. Matos, B.T. Ferreira, H. Carvalho, P. Peças, Impact assessment of additive manufacturing on sustainable business models in industry 4.0 context, Sustainability 12 (17) (2020) 1–21.
[25] W. Gao, Y. Zhang, D. Ramanujan, K. Ramani, Y. Chen, C.B. Williams, C.C.L. Wang, Y.C. Shin, S. Zhang, P.D. Zavattieri, The status, challenges, and future of additive manufacturing in engineering, Comput. Aided Des. 69 (2015) 65–89.
[26] N. Shahrubudin, T.C. Lee, R. Rohaizan, An overview on 3D printing technology: technological, materials and applications, Procedia Manuf. 35 (2019) 1286–1296.
[27] T. Femmer, A.J.C. Kuehne, M. Wessling, Print your own membrane: direct rapid prototyping of polydimethylsiloxane, Lab Chip 14 (15) (2014) 2610–2613.
[28] M. Singh, A.P. Haring, Y. Tong, E. Cesewski, E. Ball, R. Jasper, E.M. Davis, B.N. Johnson, Additive manufacturing of mechanically isotropic thin films and membranes via microextrusion 3D printing of polymer solutions, ACS Appl. Mater. Interfaces 11 (6) (2019) 6652–6661.
[29] Y. Song, B. Wang, P. Altemose, C. Kowall, L. Li, 3D-Printed membranes with a zwitterionic hydrogel coating for more robust oil-water separation, Ind. Eng. Chem. Res. 59 (48) (2020) 21058–21065.
[30] A. Vyatskikh, A. Kudo, S. Delalande, J.R. Greer, Additive manufacturing of polymer-derived titania for one-step solar water purification, Mater. Today Commun. 15 (2018) 288–293.
[31] W.S. Tan, C.K. Chua, T.H. Chong, A.G. Fane, A. Jia, 3D printing by selective laser sintering of polypropylene feed channel spacers for spiral wound membrane modules for the water industry, Virtual Phys. Prototyp. 11 (3) (2016) 151–158.
[32] J. Balster, I. Punt, D.F. Stamatialis, M. Wessling, Multi-layer spacer geometries with improved mass transport, J. Membr. Sci. 282 (1–2) (2006) 351–361.
[33] A.R. Da Costa, A.G. Fane, D.E. Wiley, Spacer characterization and pressure drop modeling in spacer-filled channels, J. Membr. Sci. 87 (1994) 79–98.
[34] L.D. Tijing, J.R.C. Dizon, I. Ibrahim, A.R.N. Nisay, H.K. Shon, R.C. Advincula, 3D printing for membrane separation, desalination and water treatment, Appl. Mater. Today 18 (2020) 100486.
[35] G. Djamel, A.E. El-Wakil, A. Abdulaziz, E. Noureddine, M. Ammar, Membrane post-synthesis modifications and how it came about, Int. J. Adv. Appl. Sci. 5 (2) (2018) 60–64.
[36] S. Hermans, E. Dom, H. Marien, G. Koeckelberghs, I.F.J. Vankelecom, Efficient synthesis of interfacially polymerized membranes for solvent resistant nanofiltration, J. Membr. Sci. 476 (2015) 356–363.

[37] O. Agboola, O.S.I. Fayomi, A. Ayoola, O.A. Augustine, E.A. Edith, E.S. Samuel, E.O. Emmanuel, M. Lucey, S. Rotimi, W.K. Kehinde, A. Babalola, A review on polymer nanocomposites and their effective applications in membranes and adsorbents for water treatment and gas separation, Membranes 11 (2021) 139.
[38] P. Sun, L. Zhang, S. Tao, Preparation of hybrid chitosan membranes by selective laser sintering for adsorption and catalysis, Mater. Des. 173 (2019) 107780.
[39] S. Mondal, S.R. Wickramasinghe, Produced water treatment by nanofiltration and reverse osmosis membranes, J. Membr. Sci. 322 (1) (2008) 162–170.
[40] K.P. Lee, T.C. Arnot, D. Mattia, A review of reverse osmosis membrane materials for desalination-Development to date and future potential, J. Membr. Sci. 370 (1–2) (2011) 1–22.
[41] P.K. Rai, J. Lee, S.K. Kailasa, E.E. Kwon, Y.F. Tsang, Y.S. Ok, K.H. Kim, A critical review of ferrate (VI)-based remediation of soil and groundwater, Environ. Res. 160 (2018) 420–448.
[42] Y. Ying, W. Ying, Q. Li, D. Meng, G. Ren, R. Yan, X. Peng, Recent advances of nanomaterial-based membrane for water purification, Appl. Mater. Today 7 (2017) 144–158.
[43] S. Munirasu, M.A. Haija, F. Banat, Use of membrane technology for oil field and refinery produced water treatment-a review, Process Saf. Environ. Protect. 100 (2016) 183–202.
[44] M. Cheryan, N. Rajagopalan, Membrane processing of oily streams. Wastewater treatment and waste reduction, J. Membr. Sci. 151 (1) (1998) 13–28.
[45] N.W.I. Syarifah, Y. Norhaniza, A. Farhana, M. Nurasyikin, A review of oilfield wastewater treatment using membrane filtration over conventional technology, Malaysian J. Anal. Sci. 21 (3) (2017) 643–658.
[46] M. Padaki, M.R. Surya, M.S. Abdullah, N. Misdan, A. Moslehyani, M.A. Kassim, N. Hilal, A.F. Ismail, Membrane technology enhancement in oil-water separation. A review, Desalination 357 (2015) 197–207.
[47] L. Dan, W. Huanting, Recent developments in reverse osmosis desalination membranes, J. Mater. Chem. 20 (2010) 4551–4566.
[48] S. Ford, M. Despeisse, Additive manufacturing and sustainability: an exploratory study of the advantages and challenges, J. Clean. Prod. 137 (2016) 1573–1587.
[49] H. Kitano, K. Takeuchi, J. Ortiz-Medina, I. Ito, A. Morelos-Gomez, R. Cruz-Silva, T. Yokokawa, M. Terrones, A. Yamaguchi, T. Hayashi, M. Endo, Enhanced desalination performance in compacted carbon-based reverse osmosis membranes, Nanoscale Adv. 2 (8) (2020) 3444–3451.
[50] T. Luukkonen, J. Yliniemi, H. Sreenivasan, K. Ohenoja, M. Finnila, G. Franchin, P. Colombo, Ag- or Cu-modified geopolymer filters for water treatment manufactured by 3D printing, direct foaming, or granulation, Sci. Rep. 10 (1) (2020) 1–14.
[51] E.P. Koumoulos, E. Gkartzou, C.A. Charitidis, Additive (nano) manufacturing perspectives: the use of nanofillers and tailored materials, Manuf. Rev. 4 (2017) 12.
[52] M. Abdalla, M. Nasser, F.A. Kayvani, H. Qiblawey, A. Benamor, S. Judd, Impact of combined oil-in-water emulsions and particulate suspensions on ceramic membrane fouling and permeability recovery, Separ. Purif. Technol. 212 (2019) 215–222.
[53] L.C. Hwa, M.B. Uday, N. Ahmad, A.M. Noor, S. Rajoo, K.B. Zakaria, Integration and fabrication of the cheap ceramic membrane through 3D printing technology, Mater. Today Commun. 15 (2018) 134–142.
[54] L. Hernandez-Afonso, R. Fernandez-Gonzalez, P. Esparza, M.E. Borges, S.D. Gonzalez, J. Canales-Vazquez, J.C. Ruiz-Morales, Ceramic-based 3D printed supports for photocatalytic treatment of wastewater, J. Chem. 2017 (2017) 9, https://doi.org/10.1155/2017/7602985. Article ID 7602985.

[55] H. Dommati, S.S. Ray, J.C. Wang, S.S. Chen, A comprehensive review of recent developments in 3D printing technique for ceramic membrane fabrication for water purification, RSC Adv. 9 (29) (2019) 16869–16883.
[56] A.V.O. Akowanou, H.E.J. Deguenon, L. Groendijk, M.P. Aina, B.K. Yao, P. Drogui, 3D-printed clay-based ceramic water filters for point-of-use water treatment applications, Prog. Addit. Manuf. 4 (3) (2019) 315–321.
[57] J. Li, R.D. Sanderson, E.P. Jacobs, Ultrasonic cleaning of nylon microfiltration membranes fouled by Kraft paper mill effluent, J. Membr. Sci. 205 (2002) 247–257.

Further reading

[1] C. Honghui, Y. Wenguang, S. Lujing, C. Shuxiang, Y. Rendi, L. Wenfeng, Y. Haibo, L. Lianqing, 4D printing: a review on recent progresses, Micromachines 11 (9) (2020) 796.

CHAPTER

7

Application of nanotechnology in water and wastewater treatment and the vast vision for the future

Sukanchan Palit[1], *Pratik Das*[2], *and Piyali Basak*[2]

[1] **Department of Chemical Engineering, University of Petroleum and Energy Studies, Dehradun, Uttarakhand, India;** [2] **School of Bioscience and Engineering, Jadavpur University, Kolkata, West Bengal, India**

7.1 Introduction

In today's society, science and technology are fast progressing toward new scientific regeneration. Environmental protection, groundwater and drinking water remediation, and industrial wastewater treatment are all on the verge of new scientific understanding and consideration. Global water scarcity, human-induced climate change, and destruction of ecological and biological diversity and integrity are urging scientists and engineers to move toward new scientific innovation and scientific ingenuity in the field of environmental engineering and environmental science. Conventional environmental engineering approaches, such as membrane science, and nonconventional environmental engineering techniques, such as advanced oxidation processes, are undergoing fresh scientific imagining and redesigning. Today's human society requires effective and cost-effective environmental engineering technologies and procedures. Today, rapid, widespread, and aggressive industrialization, mass manufacturing, and massive industrial growth are harming the environment and ecological biodiversity. In such a difficult situation, environmental scientists, researchers, students, environmental engineers and policy makers across the world are in search of new scientific innovations and engineering vision. In this treatise, the author profoundly

3D Printing Technology for Water Treatment Applications
https://doi.org/10.1016/B978-0-323-99861-1.00005-9

describes the recent advancements in drinking water and wastewater treatment. Nanomaterials and artificial nanomaterials are now being used extensively in the treatment of drinking water and industrial wastewater. A new epoch in the field of environmental engineering and chemical engineering will emerge if scientists, engineers, students, researchers, policy makers, and governments all around the world work together to reduce environmental pollution, provide safe drinking water, and treat industrial wastewater. The use of nanotechnology and nanomaterials in arsenic and heavy metal groundwater treatment and drinking water remediation is urgently needed. The author enumerates various fields of scientific and engineering research in great detail. Humankind and humanity are today in the need of new scientific and engineering vision, innovations, discoveries, and ingenuity in the domain of environmental engineering, environmental sciences, and water purification. The other areas of these scientific research endeavors are the areas of 3D (3-dimensional) printing, device development, and water and wastewater treatment. Three-dimensional printing and device development are the ingenuities of science and technology today. A newer domain of scientific ingenuity and scientific vision in the field of 3D printing and device development will surely evolve if environmental engineers and environmental scientists pursue positive research and development endeavors in the right futuristic direction.

7.2 The vision of this study

Drinking water difficulties are now afflicting human society and the human species. In the same manner, environmental protection issues are undergoing extensive vision and contemplation. Rapid industrial growth, as well as advances in manufacturing and mechanical engineering, is currently damaging the environment and ecology. Humanity is currently on the lookout for fresh scientific advancements in science and engineering. Because of the massive damage of the environment and ecology, there is an urgent need for a complete treatise on drinking water and industrial wastewater treatment. Thus, the goal and purpose of this research are to shed light on breakthroughs in the fields of drinking water treatment and industrial wastewater treatment. The author addresses contemporary advances in conventional and nonconventional environmental engineering techniques, novel separation processes such as membrane science, and the use of nanotechnology and nanoscience in environmental pollution control. Water scarcity, global warming, and climate resilience are wreaking havoc on the worldwide scientific foundation. Thus, the purpose of this book is to shed light on the use of nanomaterials and engineered nanomaterials in the treatment of drinking water and industrial wastewater. Petroleum engineering, chemical engineering, and environmental engineering science are currently at the crossroads of scientific inquiry and immense scientific prowess. Today, the loss of fossil fuel supplies is a burden on humankind. If scientists and engineers on the global platform focus on these dire environmental protection challenges, a new scientific regeneration, and scientific rejuvenation will undoubtedly begin. The author deals with these very scientifically validated challenges in the field of nanotechnology and nanomaterials use. Other areas of research and development mentioned in this well-researched article include 3D (three-dimensional) printing and applications in environmental pollution control and water remediation.

7.3 The need and the rationale of this study

Science and technology are advancing at a breakneck pace in today's global scene. In a similar vein, environmental engineering, chemical engineering, material science, and nanotechnology are also undergoing scientific revitalization. Water challenges, as well as global climate change and climate resilience issues, are wreaking havoc on the large and diverse scientific landscape. Scientific confirmation, profound scientific vision, and engineering and technical competence will all serve as torchbearers for a new scientific order in environmental research and water pollution prevention. Material science and nanotechnology/nanoengineering are diametrically opposed fields. In the human race, there is a huge demand for smart materials and ecomaterials applications. There is also a need for the use of nanomaterials in water treatment as well as environmental and industrial pollution management. The knowledge prowess of environmental sciences and engineering, the needs of human society, and the futuristic vision of technology and science will all serve as pallbearers for a newer scientific order in the fields of environmental pollution control, industrial pollution control, and water treatment. The author discusses the scientific and engineering challenges of using nanotechnology in drinking water and groundwater clean-up.

7.4 The scientific doctrine in the field of nanomaterials and engineered nanomaterials and the need of sustainability

Nanomaterials and engineered nanomaterial applications across a wide range of science and engineering fields are the true vision of scientific progress today. As human culture, science, and engineering progress, scientific theory in the fields of nanomaterials and engineered nanomaterials must be reviewed and envisioned. Nanotechnology applications in several fields of science and technology are currently on the verge of a new scientific renaissance. Environmental sustainability, water sustainability, and industrial pollution control are now inextricably connected. The ultimate necessity of the hour is a thorough introspection in the realm of green sustainability. Today's drinking water treatment tools are at the crossroads of scientific profundity and vision. Today, the world of science and technology is in shambles as arsenic groundwater poisoning wreaks havoc in many developing, disadvantaged, and developed countries around the world. Nanotechnology is also being used to remediate groundwater and drinking water. The goal of this well-researched paper is to shed light on the use of nanotechnology and other sophisticated treatment techniques in the treatment of industrial wastewater.

7.5 What exactly do we mean when we say nanoparticles and designed nanomaterials?

Nanomaterials and smart materials are scientific and technological marvels in the worldwide scientific firmament. Nanomaterials are defined as materials with at least one exterior dimension measuring 1–100 nm. According to the European Commission's definition, at least half of the particles in the number size distribution must have a particle size of 100 nm or less. Nanomaterials research and vision adopt a material science and nano-science-based approach to nanotechnology, comprising materials metrology and synthesis incorporating microfabrication. They can be optical, electrical, thermophysical, or mechanical in nature. In the realms of research and development, there is a new scientific regeneration and renewal. Humans have considerably created and manufactured artificial nanoparticles to have specific qualities and properties. Carbon black and titanium dioxide nanoparticles are two examples. In this article, the author delves into the fields of nanotechnology applications in water and wastewater treatment tools, environmental and water sustainability challenges linked with it, and the author's broad vision for the future.

7.6 Environmental sustainability, scientific progress, and a broad vision for the future

In the global context, sustainable development is an absolute necessity. Sustainable development, whether it is energy, environmental, social, or economic, will go a long way toward ensuring science and technology's true emancipation in the global scene. In many countries around the world, essential human social requirements and necessities such as drinking water, food, sanitation, education, human environment, and housing are severely ignored. As a result, the necessity for energy and environmental sustainability in human scientific and technological growth. Sustainable development, whether energy or environmental, is rich with vision and technological motivation in today's culture. As technology, engineering, and humanity advance, the wide visionary future in the field of water and industrial wastewater treatment must be reviewed and rethought.

7.7 Recent scientific advances in environmental preservation, as well as a visionary road forward

With the advancement of research and engineering in the global landscape, environmental science, environmental engineering, and environmental protection must be envisioned and reviewed. The author explains the advances in the realm of potable water and wastewater treatment in this part. This section's scientific study focuses on conventional and nonconventional environmental science and engineering methodologies. Other scientific endeavors in this field include novel separation methods, membrane separation processes, and desalination. Global water scarcity is wreaking havoc in many water-stressed areas throughout the world. As a result, a scientific and technological vision is one of the most pressing demands of the hour.

Hashim et al. [1] deeply elaborated with scientific vision and cogent insight remediation technologies for recalcitrant heavy metal contaminated groundwater [1]. Heavy metal pollution of groundwater, whether from natural or anthropogenic sources, is a major scientific and engineering concern for citizens, public health, and the enormous realm of public health engineering worldwide. Remediation and restoration of polluted groundwater are top priorities since billions of people throughout the world rely on it for safe drinking water [1]. This article reviews and categorizes 35 options for groundwater and drinking water treatment techniques into three broad categories: chemical, biochemical/biological/biosorption, and physicochemical treatments. At the end of each procedure, comparison tables have been developed and researched in order to gain a much broader understanding of each category of the environmental engineering tools presented. Keeping in mind the extensive environmental sustainability difficulties, public health engineering issues, and environmental health ethics, sophisticated procedures including natural chemistry, bioremediation, biorestoration, and biosorption are strongly advised to be used in appropriate instances [1]. The other areas of research pursuit are chelate extraction and chemical washings [1]. Pollutants can become mobile in soils depending on the pH and properties of the soil. Heavy metal toxicity and pollution can occur as a result of contaminated drinking water, food chain consumption, or excessive air concentrations near emission sources [1]. Vast and varied scientific contemplation and introspection in the field of heavy metal decontamination of drinking water are the immediate needs of the hour. Over the past few years, many water remediation and water treatment technologies were applied all over the world to deal with contaminated soil and aquifers. Thus, a new scientific understanding and a new discernment of engineering and technology are the scientific imperatives of the moment [1]. The authors deeply discussed with scientific vision and insight heavy metals in groundwater and its sources, chemical property and speciation, chemical treatment technologies, in-situ treatment by using reductants, reduction by dithionite, reduction by gaseous hydrogen sulfide, reduction by using iron-based technologies, in-situ flushing, in-situ chemical fixation, biological and biosorption treatment technologies. Enhanced biorestoration and biosorption of heavy metals are the areas of scientific research endeavors of this well-researched treatise. Physicochemical treatment technologies such as permeable reactive barriers, chemical precipitation in permeable reactive barriers, and alkaline complexation agents in permeable reactive barriers are the other pivots and cornerstones of this research pursuit. Adsorption, filtration, separation, and absorption mechanisms are the scientific and engineering needs of scientific endeavors and scientific progress in environmental engineering and chemical engineering [1]. Groundwater treatment technologies and advances have come a long way since the days of its scientific and engineering inception and beginning. A remarkable epoch in the field of groundwater and drinking water treatment will surely evolve if concerted efforts are taken by scientists, engineers, civil society, policy experts, and governments across the globe [1].According to some researchers, choosing a site-specific soil remediation technique can be a difficult and daunting research endeavor due to the uncertainty of contamination levels, the high costs of remediation, and the scientific impacts of the techniques on green biodiversity, ecology, and the environment. Biological or biochemical procedures that use bacteria and nutrients have emerged as the most promising technology in the recent decade. According to the authors of this extensive study endeavor, due to extremely complicated soil chemistry, broad and rigorous site-specific research is required to bring out the best performance from any of these technologies [1]. A new scientific order in the field of environmental remediation will undoubtedly develop in the scientific landscape if good scientific research pursuit focuses on vision, candor, scientific and engineering flair, and validation [1].

Gogate and Pandit [2] deeply discussed and elaborated in details in a review paper important technologies for wastewater treatment and the chemical oxidation technologies at ambient conditions. Because of the increasing abundance of molecules that are resistant to microorganisms in wastewater streams, conventional and traditional biological approaches cannot be used for comprehensive effluent treatment. The development of new technologies and discoveries to degrade these persistent compounds into smaller molecules that can then be destroyed by biological mechanisms has become critical [2]. Because of the increasing abundance of molecules that are resistant to microorganisms in wastewater streams, conventional and traditional biological approaches cannot be used for comprehensive effluent treatment. The development of new technologies and discoveries to degrade these persistent compounds into smaller molecules that can then be destroyed by biological mechanisms has become critical [2]. This paper delves into the fundamentals of each separate process, such as optimal operating parameters and reactor design concepts. New scientific developments and inventions in numerous disciplines of research are broadly opening new doors in the fields of environmental remediation, water sustainability, water and wastewater treatment, and so on. Waste minimization, waste valorization, zero-waste rules, and constraints are the scientific and engineering order of the day in today's scientific arena. It is critical today to reduce the amount of recalcitrant compounds in the effluent stream to a safe level in order to meet environmental regulations. The oxidation technologies discussed in this work can be divided into advanced oxidation processes (cavitation, photocatalytic oxidation, and Fenton's chemistry) and chemical oxidation processes (use of ozone and hydrogen peroxide), and these processes have enormous potential to partially or completely degrade these toxic chemicals, biorefractory chemicals, pesticides, and so on [2]. Advanced oxidation procedures and techniques are roughly characterized as processes that generate significant amounts of hydroxyl radicals in order to successfully oxidize the bulk of the compounds contained in effluent water. As human civilization and scientific advancement advance, a new scientific and technological landscape in the field of environmental and water remediation is fast emerging [2]. Cavitation, acoustic cavitation, hydrodynamic cavitation, photocatalysts, and an investigation of work done in recent years, as well as a large scope for future study, are areas of scientific pursuit. Advanced oxidation processes such as cavitation, photocatalytic oxidation, and Fenton chemistry work on the principle of free radical formation and subsequent attack on contaminant molecules, whereas ozonation and hydrogen peroxide work on the premise of direct oxidant attack. This dissertation thoroughly covers the scientific potential, scientific validation, and scientific innovation in the subject of resistant chemical pollution control [2]. The design of reactors is scientifically significant and requires a scientific vision in chemical engineering and environmental engineering systems. It is a significant aspect in influencing ozonation process efficiency and should aim to maximize contact between ozone gas and liquid effluent. Overall, photocatalytic oxidation is by far the greatest technique, whereas cavitation is a less efficient process. Scientific and engineering efforts are now necessary on the experimental, conceptual, and theoretical fronts before large-scale application may be attempted [2].

Mitigating arsenic drinking water contamination is immensely needed at this crucial juncture of civilization, science, engineering, and technology. Bangladesh and the state of West Bengal, India, are in the throes of civilization's largest environmental catastrophe. Ref. [3] deeply discussed and deliberated with vision, insight and scientific forbearance poisoning and risk assessment of arsenic in groundwater. The author intensely discussed few alarming points like (1) arsenic poisoning through ages and the victims of this immense venom, (2) the global scenario of arsenic groundwater catastrophe, (3) groundwater arsenic discontinuity and spatial mapping, spatial planning and public participation, (4) an environmental health concern due to chronic arsenic exposure to drinking water, (5) risk from groundwater arsenic exposure and epidemiological and spatial assessment, (6)arsenic induced health and social hazard and survival strategies of arsenicosis patients, (7) policy response and arsenic mitigation in Bangladesh, (8) arsenic poisoning in Bangladesh and legal issues of responsibility and the way forward and the vast vision for the future [3]. Arsenic is a proven carcinogen, and even trace amounts can pose a serious health risk. Man's vast vision and academic rigor in groundwater and drinking water treatment, as well as the futuristic vision of environmental engineering science, will undoubtedly advance civilization and mankind to newer scientific and engineering frontiers [3]. The primary scientific and technological findings of arsenic groundwater decontamination, as well as its broad scientific vision and scientific insight are now the forerunners of a newer engineering and technology order. Mankind's technological achievement will undoubtedly be bolstered if underdeveloped countries, developed countries, and emerging economies emerge from this catastrophic calamity [3]. Deep scientific provenance, the wide vision of conventional and nonconventional environmental engineering methods and tools will surely open new doors of scientific challenges, scientific potential, and scientific intuition in years to come [3]. Decontamination of arsenic groundwater is a massive issue to humankind and scientific development today. The author emphasizes the importance of numerous fields of science and engineering, such as chemical process engineering, geological sciences, environmental engineering, metallurgical engineering, and material science, in the implementation and execution of environmental remediation strategies

[3]. It is highly ironic that so many tube wells have been installed for drinking water in Bangladesh that are safe from waterborne diseases but contaminated with high levels of arsenic [3]. This contamination caused widespread environmental and health issues. More than 300 million people in 70 countries are thought to be at high risk of arsenic poisoning from groundwater [3]. Apart from Bangladesh and the neighboring Indian state of West Bengal, which between them have the largest problem, there have been serious warnings from Argentina, Chile, Taiwan, Vietnam, China, Pakistan, Thailand, and even the southwestern part of United States of America [3]. This well-researched chapter focuses on methodological issues of spatial, quantitative, and qualitative vision and investigation on arsenic poisoning, such as using Geographical Information System (GIS) to vastly investigate the distribution of arsenic-laced water in space time to uncover and reveal the distribution of water at the subsurface level [3].

Ref. [4] discussed in detail about the insight nanomaterials for industrial water and wastewater treatment procedures. The treatment of industrial wastewater is a universal scientific, technological, and engineering concern. Many persistent organic contaminants, heavy metals, and recalcitrant chemicals are present in high amounts in wastewater [4]. To remove these pollutants, effective environmental engineering processes and scientific innovation are necessary. Nanotechnology has applications in a wide range of human scientific fields, including applied sciences, nanoengineering, chemical process engineering, environmental engineering, geological sciences, material science, and composite science. This book delves closely into the utilization of nanoparticles for the removal of various pollutant materials from industrial wastewater, with a particular emphasis on metals and metal oxide nanomaterials (NMs), carbon nanomaterials, and nanofiber/nanocomposite membranes [4]. The other pivots of this work are emerging nanomaterials and their strong scientific evidence and scientific attitude [4]. The authors covered industrial wastewater sources and compositions, industrial wastewater treatment procedures, metal and metal oxide nanoparticles, silver nanomaterials, iron nanoparticles, titanium dioxide nanoparticles, and iron oxide nanoparticles with scientific rigor [4]. Other highlights of this paper include carbon-based nano adsorbents and their applications. Nanomaterials have a number of distinct physicochemical properties. These characteristics make them extremely appealing for wastewater treatment: (1) greater surface areas compared to typical nanomaterials, (2) capacity to be functionalized, (3) employment as high selectivity recyclable zones for harmful components in effluent streams, (4) a broad future vision [4]. The chemical kinetics and absorption mechanisms of various adsorbents are significantly dependent on the adsorbed materials, as well as physicochemical characteristics such as solution pH, original concentration of the resistant dye or metal, dose, and reaction temperature. The reuse and recycling of industrial wastewater and groundwater have enormous scientific potential in the route of human scientific and technical progress. As a result, the writers focus on these scientific and engineering difficulties and pressing issues [4].

Environmental remediation, water science and engineering, and water pollution control are all major challenges in today's global scientific scenario. The use of conventional and nonconventional environmental techniques, as well as membrane research and desalination, is critical. Water-stressed countries around the world are now pursuing new tools in desalination research. The use of nanomaterials in water and wastewater treatment is one of the most recent and unique areas of scientific understanding and vision. The author forcefully proclaims these topics of exciting research inquiry throughout the entire chapter.

7.8 Recent scientific advances in the application of nanotechnology in water and wastewater treatment

In the global context, nanotechnology, nanoengineering, and nanoscience applications in water and wastewater treatment are new fields of scientific investigation. Nanomaterials and artificial nanomaterials are emerging areas of scientific inquiry that are being studied with scientific and academic rigor. A scientific understanding of nanotechnology's application in drinking water and groundwater remediation will surely be the forerunners of a new scientific and engineering order on a global scale. In this section, the author discusses recent scientific and engineering advancements in the disciplines of nanotechnology and water and wastewater treatment. As a result, the world of science and technology will definitely be riveted as governments and legislators all over the world take swift action to address global water scarcity and human-caused climate change.

Madhura et al. [5] deeply elaborated on nanotechnology-based water quality management for potable, handy and industrial wastewater treatment. The most difficult problem is to create innovative and sustainable wastewater treatment technologies that are low in production cost and high in efficiency. This study focuses on cutting-edge treatment tools based on nanomaterials such as graphene and carbon nanotubes, metal nanoparticles such as silver, iron, zinc, and magnesium, metal oxide nanomaterials, and magnetic core composites containing cobalt, iron, and nickel [6,7]. In addition, the authors examined developing and established methods in terms of cheap production costs and successful and long-term removal of organic and inorganic contaminants. The authors explored nanotechnology-based water treatment, as well as water reuse and recycling techniques, as well as the urgent challenges of nanotechnology's function in water treatment [6,7].

Mansoori et al. [8] described with scientific conscience and scientific girth environmental application of nanotechnology. Nanotechnology is a new and promising sector of science and technology. It plays a significant and positive part in the development of inventive and original approaches for producing newer methods and goods to replace old industrial equipment and inventing new materials, chemicals, and technologies [8]. In the global scenario, scientific validity, scientific flair, and deep scientific provenance are the most pressing needs of science, technology, and engineering. Environmental applications of nanotechnology are significantly concerned with the creation of answers and methods to existing environmental problems as well as preventive measures for future energy and material difficulties [8]. This chapter provides a comprehensive and extensive summary of the expanding research and development projects on environmental remediation with nanotechnology. The key and crucial aspects of environmental concerns are first thoroughly addressed, followed by a description of the applications of nanotechnology to substances that can be regarded as environmental rehabilitation [8]. Nanotechnology is now being used in water resource management and wastewater management. As environmental preservation and water remediation usher in a new era, humanity's great scientific prowess, academic rigor, and scientific rigor are bolstered. Titanium dioxide, iron, bimetallics, catalytic particles, clays, carbon nanotubes, fullerenes, dendrimers, and magnetic nanoparticles are among the nanoparticles researched [8]. Nanotechnology is a subcategory of technology in colloidal science, biology, physics, chemistry, material science, and other scientific fields. Similar to the success of nanotechnology nanoscale materials and smart materials have the potential to improve the environment in consumer products and other sectors, both directly through applications of these smart and ecomaterials to prevent and remove recalcitrant pollutants and indirectly through the use of nanotechnology to design cleaner and more environmentally friendly processes [8]. In this chapter, the writers go into great detail about titanium dioxide-based nanoparticles, iron-based nanoparticles, bimetallic nanoparticles, and nanoparticulate photocatalysts, and other catalysts. This book is distinguished by nano clays, nanotubes, dendrimers, and nano sponges. The other pivots of this research effort include nanomembranes and nano sieves, as well as their properties and applications [8,9]. If scientists, researchers, and engineers throughout the world focus on newer discoveries in the areas of water shortage, climate change, and water remediation, a new dawn in the field of nanoscience, nanotechnology, and nanoengineering will undoubtedly come in near future [8,9].

Konduru et al. [10] discussed with cogent insight nanotechnology for water purification and applications of nanotechnology methods in sustainable industrial wastewater treatment and wastewater and water sustainability. Today, a new scientific breakthrough is taking place in the realm of modern and sustainable water and wastewater treatment systems. Water is the most valuable asset of human civilization, and access to safe drinking water is a basic human requirement. Population expansion, global climate change, and deterioration of water quality all contribute to an increase in demand for fresh water [11]. Humankind's scientific acumen and profundity in the fields of water treatment and environmental remediation are seeing a renaissance [11]. The concept of environmental engineering science and chemical engineering is currently in the midst of a worsening catastrophe in several South Asian countries due to arsenic and heavy metal groundwater poisoning. As a result, the scientific and engineering orders of humanity are to reuse, recycle, and reduce [11]. Water pollutants are poisonous and carcinogenic, wreaking havoc on persons, public health, and ecosystems. Some heavy metals, such as arsenic, are well-known harmful contaminants. Cadmium, mercury, lead, zinc, copper, mercury, nickel, and other heavy metal contaminants with severe toxicity are also present. Pesticides, fertilizers, hydrocarbons, phenols, plasticizers, biphenyls, detergents, and greases are all connected with significant levels of toxicity. Pharmaceuticals, medicines, and personal care items are examples of growing pollutants in today's human progress [11]. The authors thoroughly covered the importance of nanotechnology in water purification, significant constraints and obstacles in water purification, nano adsorption, an overview of diverse nanomaterials uses in water and wastewater treatment, and carbon nano adsorbent applications. Circular economics and environmental engineering research are now linked by an umbilical cord. As a result, a thorough understanding of water and wastewater treatment is required [11]. The other scientific and technological foundations of this dissertation include membranes and membrane processes, as well as nanocomposite membranes. Other topics of scientific investigation include photocatalysis and biologically inspired membranes. The other pivots of this article are water purification techniques based on nanopolymers [11]. Other key aspects of this treatise are the safety, toxicity, and environmental impact of nanoparticles [11]. Today, there is a greater need for sustainable air and water treatment technologies that provide high-quality drinking water, improved air quality, the removal of micropollutants, and the intensification and control of industrial processes. Other demands of humanity and human science and engineering advancement today are sustainable environments. The authors of this paper discussed the fascinating scientific and engineering concerns of water and industrial wastewater treatment [11]. Today, environmental, energy, and nanotechnology sustainability are all intertwined. The buzzwords of today's scientific and engineering progress are nanotechnology, nanobiotechnology, and environmental biotechnology. The author went into great detail about scientific creativity, forbearance, and scientific and engineering vision.

7.9 Recent scientific advancements in the application of environmental sustainability in human society and the road ahead

The application of environmental sustainability in humanity and human scientific advancement is a visionary torchbearer for a new scientific and engineering order in environmental and industrial pollution management, as well as water and industrial wastewater treatment. Green sustainability is the real vision of science and engineering on the global scientific and engineering platform. Mankind's extensive and flexible scientific understanding, man's scientific and academic rigor, and science and technology's futuristic vision will all be torchbearers for global sustainability challenges. The author clarifies and validates these scientific and technological difficulties in this part.

Ref. [12] discussed with scientific and technological vision environmental/green sustainability and its research growth and trends [12]. The research being conducted in the field of sustainability is growing, as is the domain of environmental management, environmental integrity, water sustainability, and sustainable development. The goal of this paper is to outline the research trends, define the literature characterization, and research forays of environmental sustainability engineering research from the perspective of historical evaluation based on the top five highest impact factor journals in the Institute of Scientific Information's Journal Citation Report [10]. As civilization, science, and technology advance, scientific study in the realm of environmental and green sustainability must be rethought and re-envisioned [12]. The United Nations World Commission on Environment and Development defined "sustainable development" as "development that meets the demands of the present without jeopardising future generations' ability to satisfy their own needs." The authors thoroughly discussed (1) research trends in sustainability research, (2) literature categorization of sustainability research, (3) sustainability research focus, and (4) a detailed discussion of future research trends and the flow of scientific thoughts and engineering vision in sustainability [12]. Research in sustainability, whether environmental, energy, social, or economic, has advanced dramatically during the last few decades [12]. Water and environmental protection research has been a persistent focus of sustainability research throughout the previous few decades. However, progress in green manufacturing/green engineering research is modest [12]. Survey research in sustainability can provide a more in-depth scientific understanding of the main concerns and difficulties that researchers encounter [10]. The authors believe that this study can help academics decide where to focus their research because the findings provide actual information about sustainability demands, scientific potential, and challenges. In the coming years, a thorough scientific understanding, scientific knowledge, and scientific know-how in the field of sustainable development will open up new avenues for innovation and scientific vision [12].

The United States Environmental Protection Agency Report (2011) delves further into sustainability and the United States Environmental Protection Agency. Today, the buzzwords of human scientific advancement are environmental sustainable development, environmental chemical engineering, and environmental management. Recognizing the relevance of sustainability in its work, the US Environmental Protection Agency has been investigating applications in a variety of fields in order to better incorporate sustainability into decision making [13]. The report goes into detail about the history of sustainability, a sustainability framework for the EPA, sustainability assessment and management, and risk management and reassessment principles. The benefits of a sustainability strategy in the running of the Environmental Protection Agency include additional areas of deep scientific introspection and scientific vision [13]. The growing recognition of sustainability as both a process and a goal for long-term human well-being, economic growth, and advancement is founded on four essential concepts. The first is an acknowledgment that present measures targeted at risk reduction are ineffective. Second, sophisticated tools are available to address cross-cutting, complicated, and difficult challenges that are beyond the scope of the current method. Third, international organizations are using sustainability as a veritable shared approach to address the three sustainability pillars, which are social, environmental, energy, and economic challenges. Finally, the economic significance of sustainability to the United States is acknowledged [13]. If scientists, engineers, civic society, and nations around the world take bold efforts toward effective environmental or green sustainability implementation, a new window of scientific invention and scientific instinct will undoubtedly open. This report delves deeply into the scientific demands, scientific potential, and scientific advancement in the subject of environmental protection and sustainability [13].

The Environmental Protection Agency, Ireland Report [14] delves closely into the environment and health and wellness, climate change, water quality restoration and protection, and the status of nature and wild places. Today, environmental sustainability and management are critical components of government policy. Sustainable development and environmental management are diametrically opposed scientific and technological concepts [14]. This is a case study of the Irish government. Climate change, resilience, water sustainability, water quality, and waste management are today's needs of human society and the human race. This report focuses on three critical areas:

1. Informing policy development and implementation.
2. Envisioning green innovation.
3. Enhancing research capacity.

A new scientific and engineering order will undoubtedly be established if nations all over the world take bold and positive measures toward the effective implementation of environmental sustainability. The preservation of ecological biodiversity will also play a significant and crucial role in the advancement of a nation and humanity. This paper also supports the role of science, engineering, research, and scientific innovation in responding to environmental concerns and obstacles [14]. Capacity and capacity building are the report's other topics of in-depth scientific investigation. Climate change is widely regarded as a serious environmental concern in today's scientific landscape and firmament. These are the areas of serious scientific thought in this well-researched treatise.

The United States Environmental Protection Agency Report [15] explored sustainability and sustainable and effective approaches for building water utility and water needs roadmaps with vision and cogent insight. This report provides a thorough understanding of (1) utility business planning, (2) product quality, (3) customer satisfaction, (4) leadership development, (4) financial viability, (5) infrastructure stability, (6) operational resilience, (7) water resource adequacy, (8) community sustainability, and (9) performance measurement. The torchbearers of a newer world order in the sphere of environmental, natural resource, and green sustainability are scientific and engineering vision and position, the necessities of sustainability in human scientific advancement, and the futuristic vision of human civilization [16]. Sustainable drinking water and industrial wastewater services are critical principles for the advancement of the United States as a nation. As a result, there is a need for and rationale for this report.

Today, humanity, technology, and science are in the midst of patience and ingenuity. The author exhaustively enumerates the success of environmental remediation in developing a new engineering vision and economic order in the global landscape throughout this entire chapter.

7.10 3D printing, device development, and water treatment applications

Three-dimensional (3-D) printing and device development are now in use with water and wastewater applications. Three-dimensional printing science and engineering, as well as gadget creation, are currently on the verge of scientific rejuvenation and reenvisioning. This is an exciting new area of scientific investigation. In the coming years, the scientific and academic rigor in the field of 3D printing and device creation will undoubtedly open new doors of scientific vision and scientific profundity. Three-dimensional printing, also known as additive manufacturing, is the process of creating and replicating three-dimensional objects from CAD (Computer Aided Design) models or digital 3D models. The term "3D printing" refers to a wide range of procedures in which material is deposited, combined, or solidified under computer control to produce a three-dimensional item. In today's worldwide scientific scenario, 3D printing has a wide range of applications in all fields of scientific and engineering endeavor. In the 1980s, 3D printing techniques were deemed highly suited primarily for the manufacture of functional prototypes, and rapid prototyping was a more appropriate item for it at the time. In 2019, 3D printing precision, repeatability, scientific ingenuity, and material range have improved to the point that some 3D printing techniques are regarded practicable and viable as an industrial production technology. One of the primary benefits of 3D printing is the ability to construct extremely complicated forms and geometries. As humanity progresses, the vision and cleverness of 3D printing and device creation are now on the route to fresh scientific regeneration and reenvisioning. The futuristic term *additive manufacturing* exploded in popularity in the 2000s, owing to the idea of materials being added together in a variety of ways. By the early 2010s, the phrases 3D printing and additive manufacturing had expanded enormously in terms of sensibility, vision, and ingenuity, and they became substitute umbrella titles for additive technologies. Until recently, the phrase 3D printing was linked with low-cost, low-capability devices. Raymond Jones initially proposed the fundamental notion of the method to be utilized in 3D printing in his story "Tools of the Trade," which was published and envisioned in the November 1950 issue of *Astounding Science Fiction* magazine. In that paper, he referred to it as a "molecular spray." Johannes F Gottwald patented the Liquid Metal Recorder in 1971, which is a continuous inkjet metal material technology that forms a removable metal fabrication on a reusable surface. Thus, in the sphere of 3D printing and additive manufacturing, a great scientific generation and scientific profundity developed. Three-dimensional printing applications include (1) the food industry, (2) the fashion industry, (3) sustainable manufacturing, (4) environmental pollution management, water and wastewater treatment, (5) air quality monitoring, (6) the transportation industry, (7) the safety industry, and (8) the health sector. Three-dimensional printing is a method and procedure for creating three-dimensional objects by connecting materials layer by layer under computer control [17]. [14]. In the 1980s, the advent of computer-assisted 3D modeling technologies aided the creation of 3D printing. Ref. [17] discussed 3D printing approaches for environmental applications in minute detail and with scientific vision and intelligence [17]. As 3D printing technology surpassed one visionary boundary after another, a remarkable aeon in the area of science and technology developed. The authors [17] [14] went into great length about applications such as sustainable industrial manufacturing, air quality, water and wastewater treatment and reuse, and alternative energy sources. Renewable energy, energy sustainability, and comprehensive sustainable development in human civilization will open up new avenues for research and development projects in the coming years [17].

Ref. [18] carefully considered a three-dimensional printed water filtering system for onsite arsenic and heavy metals removal and decontamination with vision, scientific grit, and scientific perseverance. The perplexing threat of arsenic pollution of drinking water to mankind, civil engineering, and public health engineering, especially in developing nations, has become a severe environmental hazard. As a result, there is an urgent need for profound scientific breakthroughs. Millions of people worldwide continue to rely on groundwater as a source of drinking water, and it includes high amounts of arsenic, resulting in excessive exposure to this harmful element due to expensive costs and a lack of suitable public health engineering infrastructure [18]. As a result, there is a technique for large-scale treatment of water in rural regions to remove arsenic that is low-cost, easily tailored, ecologically friendly, and does not rely on electricity. To meet these specifications, the authors developed a three-dimensional (3D) printed water filtration and purification system for arsenic removal. The authors explored the effects of regulated surface area on flow rate and adsorbent deposition with strong scientific conscience and vision. The ambition of scientific research pursuit in the field of 3D printing devices is now greatly bolstered as environmental engineering and chemical engineering advance toward a futuristic frontier. The author validates these scientific and engineering concerns with eyesight and lucid far-sightedness in this study [18].

Today, innovative technologies such as 3D printing, device creation, and nanotechnology are reshaping the enormous scientific fabric. As environmental engineering science, chemical engineering, civil engineering, material science, and nanotechnology march forward toward a utopian future, a new scientific and engineering domain and order are being rejuvenated.

7.11 Heavy metal and arsenic groundwater remediation, the application of nanotechnology and the visionary future

The use of nanotechnology and nanomaterials in potable drinking water and industrial wastewater treatment is one of the most pressing demands of the day. Heavy metal and arsenic groundwater cleanup are perplexing environmental engineering issues in many developing, emerging, and developed countries throughout the world [19]. Arsenic pollution in drinking water is the world's most serious scientific and engineering problem, notably in South Asian countries such as India and Bangladesh. A thorough understanding of environmental engineering and nanotechnology is an urgent requirement. A sustainable society is also essential in today's world. Thus, environmentally sound and sustainable engineering techniques will undoubtedly pave the way for new scientific invention and scientific profundity in the field of environmental and water remediation. The history of science and technology for environmental protection will undoubtedly be revisited if environmental scientists and engineers take giant and effective steps in unraveling science's intricate problems. The use of nanotechnology tools and nanomaterials in groundwater remediation, as well as its scientific prowess and originality, will undoubtedly go a long way toward realizing the true potential of environmental engineering, chemical engineering, and nanotechnology.

Shannon et al. [20] deeply elucidated with scientific insight science and technology for water purification and contaminant removal issues during the next few decades. The importance of clean drinking water is well known: 1.2 billion people lack safe drinking water, 2.6 billion people lack sanitation, and millions of people die each year—3900 children every day from incurable diseases [19]. As a result, humankind is in the midst of a swift, perplexing, and ever-expanding calamity. The writers discussed wastewater disinfection, decontamination, reuse, and reclamation. The various membrane separation tools are thoroughly explained. Heavy metal removal and deterioration are issues for environmental engineers. As man and humanity progress, public health engineering will need to be reframed [21,22,23,24–27].

Ref. [28] deeply with clear insight described a thorough and detailed analysis of technology alternatives for arsenic-rich groundwater decontamination. Various physicochemical and biological methods, such as oxidation, coagulation-flocculation, adsorption, biological sorption, ion-exchange, membrane separation processes, treatment with biological organisms, and electrocoagulation, have been extensively researched in recent years by researchers all over the world. This page thoroughly examines the many technologies available for arsenic cleanup, as well as the vast removal mechanisms, cost-effectiveness, and environmental sustainability. The authors went into great detail into microbial oxidation, adsorption, and the use of red mud and activated alumina [28].

Membrane science is now considered the holy grail of industrial wastewater treatment. If governments and policy makers take good measures toward greater research and development methodologies, a new visionary age in the field of arsenic groundwater remediation will emerge.

7.12 Future scientific recommendations and future flow of scientific ideas

In terms of environmental protection/remediation and ecological biodiversity protection, humankind's future looks bleak [28,29]. Water scarcity and global climate change are causing problems in countries all over the world. Future scientific suggestions and the flow of scientific ideas should focus on standard and nontraditional environmental engineering procedures, as well as novel separation technologies such as membrane science [30–33]. As humanity, the human race, and science progress, environmental engineering and chemical engineering curricula must be envisioned in the global context [30–34]. Waste management, waste valorization, and waste minimization are all difficult and pressing challenges in today's world. Zero waste tools and zero emission standards are today's most pressing needs. As a result, the future vision of environmental engineering, chemical engineering, and nanotechnology should be oriented toward increased realization and execution of scientific grit and scientific endurance [30–38]. If scientists, engineers, governments, policy makers, and civil society take significant efforts toward environmental restoration, a new scientific and engineering order will undoubtedly develop in the global scientific fabric. Today, 3D printing and gadget creation are emerging as new fields of deep scientific understanding. They are used in the treatment of drinking water and industrial effluent. Population development places enormous demand on local sources of fresh water due to growing use and water contamination caused by industrial wastes and deteriorating urban and public health engineering systems. Thus, humanity, science, and technology are on the road to a brighter future and massive scientific regeneration [22,30–32,34–36]. Today, a wide range of water purification procedures, such as filtration, distillation, chemical purification, chemical separation tools, and other unit activities in chemical engineering, is successfully used [29,39,40]. As a result, 3D printing has numerous uses in water purification and environmental remediation [29,39–56]. The circular economy, blue economy, and green economy are all undergoing rapid scientific advancements. As a result, the circular economy should be integrated with sustainable resource management in poor countries around the world. These are the engineering problems confronting modern human civilization [55].

7.13 Conclusion, summary, and environmental engineering perspectives

In the global scenario, water and industrial wastewater treatment, minimization, and valorization are critical. Similarly, environmental, energy, and water sustainability are broad visions for the future. Sustainable development, whether it is energy-related, environmental-related, social-related, or economic-related, will undoubtedly contribute significantly to the true liberation of environmental protection and water remediation today. As civilization, science, and technology advance, environmental viewpoints in the global scenario must be reenvisioned and reorganized. Today, environmental, energy, social, and economic sustainability, as well as water sustainability, are in desperate need. Many developing, disadvantaged, and wealthy countries around the world are grappling with pressing environmental engineering concerns such as arsenic and heavy metal groundwater poisoning. Bangladesh and the Indian state of West Bengal are currently in the grip of an unprecedented environmental calamity caused by arsenic groundwater poisoning. As a result, scientific reflection is an urgent requirement of the hour. Applications of nanomaterials and artificial nanomaterials in drinking water and groundwater treatment are also critical. Scientific evidence and real-world case studies in potable water and wastewater treatment are also urgently required. In the global context, environmental engineering and chemical engineering curricula must be revised and revamped as soon as possible in order to address these massive water-related concerns. Today, sustainable development/water sustainability and environmental management are two opposing sides of the same scientific and engineering coin. These scientific and technological challenges are thoroughly addressed in this dissertation. In the next years, scientific determination and scientific girth will undoubtedly open new vistas of liberation in the fields of environmental remediation, water and green sustainability, and water and wastewater treatment.

References

[1] M.A. Hashim, et al., Remediation technologies for heavy metal contaminated groundwater, J. Environ. Manag. 92 (10) (2011) 2355–2388.
[2] P.R. Gogate, A.B. Pandit, A review of imperative technologies for wastewater treatment I: oxidation technologies at ambient conditions, Adv. Environ. Res. 8 (3–4) (2004) 501–551.
[3] M.M. Hassan, Arsenic in Groundwater: Poisoning and Risk Assessment, CRC Press, 2018.
[4] S.M. Abdelbasir, A.E. Shalan, An overview of nanomaterials for industrial wastewater treatment, Kor. J. Chem. Eng. 36 (8) (2019) 1209–1225.
[5] L. Madhura, S. Singh, S. Kanchi, M. Sabela, K. Bisetty, Nanotechnology-based water quality management for wastewater treatment, Environ. Chem. Lett. 17 (2018) 65–121.
[6] S. Homaeigohar, 'Water Treatment with New Nanomaterials', Water, Multidisciplinary Digital Publishing Institute, 2020, p. 1507.
[7] L. Madhura, et al., Nanotechnology-based water quality management for wastewater treatment, Environ. Chem. Lett. 17 (1) (2019) 65–121.
[8] G. Ali Mansoori, et al., Environmental application of nanotechnology, Annu. Rev. Nurs. Res. (2008) 439–493.
[9] I. Gehrke, A. Geiser, A. Somborn-Schulz, Innovations in nanotechnology for water treatment, Nanotechnol. Sci. Appl. 8 (2015) 1.
[10] K.A. Konduru, M. Nazarkovsky, S. Farah, R.P. Pawar, A. Basu, A.J. Domb, Nanotechnology for water purification: applications of nanotechnology methods in wastewater treatment, Chapter-2, in: A. Grumezescu (Ed.), Water Purification, Elsevier, Netherlands, 2017, pp. 33–74.
[11] K.R. Kunduru, et al., Nanotechnology for water purification: applications of nanotechnology methods in wastewater treatment, Water Purif. (2017) 33–74.
[12] F.A. Goni, et al., Environmental sustainability: research growth and trends, Adv. Sci. Lett. 21 (2) (2015) 192–195.
[13] N.R. Council, et al., Sustainability and the US EPA, National Academies Press, 2011.
[14] D. Brian Colgan, Environmental Protection Agency Report, Ireland, Report, Science and Sustainability: Research-Based Knowledge for Environmental Protection, 2016.
[15] United States Environmental Protection Agency Report, (2014), Moving towards sustainability: Sustainable and effective practices for creating your water utility roadmap. December, 2014.
[16] Moving toward Sustainability: Sustainable and Effective Practices for Creating Your Own Water Utility Roadmap, 2014. Available at: https://www.epa.ie/publications/research/previous-strategy-documents/EPA_STRIVE_PETR.pdf.
[17] M.N. Nadagouda, M. Ginn, V. Rastogi, A review of 3D printing techniques for environmental applications, Curr. Opin. Chem. Eng. 28 (2020) 173–178.
[18] K. Kim, et al., Three-dimensional, printed water-filtration system for economical, on-site arsenic removal, PloS One 15 (4) (2020) e0231475.
[19] M.A. Shannon, et al., Science and technology for water purification in the coming decades. Nanoscience and Technology: A Collection of Reviews from Nature Journals, 2010, pp. 337–346.
[20] Shannon M.A., Bohn P.W., Elimelech M., Georgiadis J.G., Marinas B.J., Mayes A.M., (2008). Science and technology for water purification in the coming decades. Nature. Nature Publishing Group, United Kingdom, pp-301-310.
[21] C. Barrow, Environmental Management for Sustainable Development, Routledge, 2006.
[22] P. Das, et al., Facile green synthesis of non-genotoxic, non-hemolytic organometallic silver nanoparticles using extract of crushed, wasted, and spent Humulus lupulus (hops): characterization, anti-bacterial, and anti-cancer studies, Environ. Res. 204 (2022) 111962.

[23] C.M. Hussain, Handbook of Nanomaterials for Industrial Applications, Elsevier, 2018.
[24] S. Palit, Advanced environmental engineering separation processes, Environmental analysis and application of nanotechnology: a far-reaching review, in: Advanced Environmental Analysis, 2016, pp. 377–416.
[25] S. Palit, Nanofiltration and ultrafiltration-the next generation environmental engineering tool and a vision for the future, Int. J. Chem. Tech. Res 9 (5) (2016) 848–856.
[26] S. Palit, Application of nanotechnology, nanofiltration, and drinking and wastewater treatment—a vision for the future, in: Water Purification, Elsevier, 2017, pp. 587–620.
[27] S. Palit, Recent advances in the application of engineered nanomaterials in the environment industry—a critical overview and a vision for the future. Handbook of Nanomaterials for Industrial Applications, 2018, pp. 883–893.
[28] S. Ghosh, A. Debsarkar, A. Dutta, Technology alternatives for decontamination of arsenic-rich groundwater—a critical review, Environ. Technol. Innov. 13 (2019) 277–303.
[29] M. Kaur, et al., Bionanomaterials from agricultural wastes, in: Green Nanomaterials, Springer, Singapore, 2020, pp. 243–260.
[30] C.M. Hussain, et al., Handbook of Environmental Materials Management, Springer, 2019.
[31] C.M. Hussain, Handbook of Manufacturing Applications of Nanomaterials, Elsevier, Amsterdam, 2020b.
[32] C.M. Hussain, B. Kharisov, Advanced Environmental Analysis: Applications of Nanomaterials, Vol. 1, Royal Society of Chemistry, 2016.
[33] C.M. Hussain. Handbook of Functionalized Nanomaterials for Industrial Applications, Elsevier, Amsterdam, Netherlands, 2020.
[34] C.M. Hussain, Handbook of Industrial Applications of Polymer Nanocomposites, Elsevier, Amsterdam, 2020a.
[35] L.M.T. Martinez, O.V. Kharissova, B. I. K., Handbook of Ecomaterials, Springer International Publishing, Switzerland, 2019.
[36] S. Palit, C. Hussain, Green sustainability and the application of polymer nanocomposites—a vast vision for the future, in: Handbook of Polymer Nanocomposites for Industrial Applications, Elsevier, 2021, pp. 733–747.
[37] S. Palit, C.M. Hussain, Nanomaterials for environmental engineering and energy applications. Handbook of Nanomaterials and Nanocomposites for Energy and Environmental Applications, 2020, pp. 1–24.
[38] S. Palit, C.M. Hussain, Green polymer nanocomposites, biocompatible nanopolymers, and the environmental pollution control: a far-reaching review. Handbook of Polymer and Ceramic Nanotechnology, 2021, pp. 3–23.
[39] D.K. Gupta, S. Chatterjee, Arsenic Contamination in the Environment: The Issues and Solutions, Springer, 2017.
[40] A. Mitra, S. Chatterjee, D.K. Gupta, Potential role of microbes in bioremediation of arsenic, in: Arsenic Contamination in the Environment, Springer, 2017, pp. 195–213.
[41] S. Ahmed, Bio-based Materials for Food Packaging: Green and Sustainable Advanced Packaging Materials, Springer, 2018.
[42] S. Ahuja, Lessons learned from water disasters of the world, in: Separation Science and Technology, Elsevier, 2019a, pp. 417–427.
[43] S. Ahuja, Overview: evaluating water quality to prevent future disasters, Separ. Sci. Technol. 11 (2019b) 1–12.
[44] T. Anukiruthika, et al., Characterisation of green nanomaterials, in: Green Nanomaterials, Springer, Singapore, 2020, pp. 43–79.
[45] S. Bajpai, N. Alam, P. Biswas, Present and potential water-quality challenges in India, in: Separation Science and Technology, Elsevier, 2019, pp. 85–112.

[46] B. DeCourten, A. Romney, S. Brander, The heat is on: complexities of aquatic endocrine disruption in a changing global climate, Separ. Sci. Technol. 11 (2019) 13–49.

[47] L.M.T. Martinez, O.V. Kharissova, B.I. Kharisov, Handbook of Ecomaterials, Springer, 2019.

[48] J.S. Metcalf, N.R. Souza, Cyanobacteria and their toxins, in: Separation Science and Technology, Elsevier, 2019, pp. 125–148.

[49] D.K. Nordstrom, D.G. Archer, Arsenic thermodynamic data and environmental geochemistry, in: Arsenic in Ground Water, Springer, 2003, pp. 1–25.

[50] S. Palit, Enzyme engineering and application of extremozymes: a vast vision for the future, in: Microbial Extremozymes, Elsevier, 2022, pp. 247–257.

[51] S. Palit, C.M. Hussain, Minerals and metal Industry in the global scenario and environmental sustainability, in: Sustainable Resource Management, Elsevier, 2021b, pp. 163–177.

[52] P. Pandit, T.N. Gayatri, Introduction to green nanomaterials, in: Green Nanomaterials, Springer, Singapore, 2020, pp. 1–21.

[53] Y. Shevah, Impact of persistent droughts on the quality of the Middle East water resources, in: Separation Science and Technology, Elsevier, 2019, pp. 51–84.

[54] P. Singh, S.K. Yadav, M. Kuddus, Green nanomaterials for wastewater treatment, in: Green Nanomaterials, Springer, Singapore, 2020, pp. 227–242.

[55] P.L. Smedley, Arsenic in groundwater—south and east Asia, in: Arsenic in Ground Water, Springer, 2003, pp. 179–209.

[56] K.G. Stollenwerk, Geochemical processes controlling transport of arsenic in groundwater: a review of adsorption, in: Arsenic in Ground Water, 2003, pp. 67–100.

CHAPTER

8

3D printed membranes for oil/water separation

Sreelakshmi Rajeevan[1,2], *and Soney C. George*[1]

[1] **Center for Nanoscience and Technology, Amal Jothi College of Engineering, Koovappally, Kerala, India;** [2] **APJ Abdul Kalam Technological University, CET Campus, Thiruvananthapuram, Kerala, India**

8.1 Introduction

The sustainability of all the living things on the Earth including humans, animals, plants, aquatic plants, and animals greatly depends on the availability of water. About 72% of the Earth's surface is covered by saltwater, and the remaining 28% is available only for the survival of life on Earth. But industrialization and human activities polluted natural resources of water such as rivers, lakes, ponds, groundwater, etc. Now, in the 21st century, the scarcity of clean water is a serious crisis that we face in our daily lives. The emerging pollutants present in water include chemicals and toxic metals from industries, agricultural wastes, medical wastes, sewage sludges, oil spillage, etc. [1], and removing all these contaminants is always a difficult task. Several methods have been adopted for wastewater treatment, and the membrane-based separation process has been found to be superior to most conventional techniques such as sedimentation, adsorption, filtration, bio-oxidation, etc. Because the latter mentioned techniques have repeatedly been shown to be ineffective in removing toxins from the water. Now, the technological advances in combining an adorable technique called additive manufacturing or 3-dimensional printing or 3D printing have taken the traditional membrane-based separation process to another level. Three-dimensional printed membranes are an excellent candidate for wastewater treatment. This chapter concisely discusses the different types of 3D printed membrane materials, their properties, advantages, and the effectiveness of removing oil from water.

3D Printing Technology for Water Treatment Applications
https://doi.org/10.1016/B978-0-323-99861-1.00004-7

8.1.1 Additive manufacturing or 3-dimensional printing technology

Additive manufacturing (AM) or 3D printing is an advanced manufacturing technique that paved a unique pathway for the design and production of forged objects with exceptionally homogeneous structures with excellent properties from the bottom layer to the top layer region of the material with the assistance of CAD software [2]. The designing of the model has been done with the help of a computer program and is fabricated in a layered pattern by different 3D printing techniques. It is also called rapid prototyping (RP) by creating prototypes of multifunctional applications. Generally, in RP, the model designed with the aid of CAD gets converted into a readable file called an STL file. STL file is a comprised form of a list of the triangular face that passes to the rapid prototyping system to form the model. RP analyses the readable file and converts it into a few 2-dimensional (2D) layers before 3D printing [3]. The first 3D printing technology was introduced by Hideo Kodama, a Japanese inventor by creating a polymer-based solid object with the help of an ultraviolet-assisted 3D printing method. The objects fabricated via AM possess high mechanical stability and uniform morphological pattern due to the layered manufacturing method. Most importantly, it is a cost-effective and time-saving process, and hardly complex structures can be fabricated easily by this technique [4]. Depending on the type of material, the AM techniques are categorized into seven processes and are depicted in Fig. 8.1. AM facilitates the fabrication of materials in different forms, such as foams, scaffolds, membranes, etc. Semipermeable membranes are an ideal candidate for water treatment. The most attractive feature of 3D printing is that each step of the process such as designing, architecture, and production of the object can be tailored and controlled as per the requirement or final application [6]. The diversity in the pore shapes and pore sizes of the membrane increased the application of 3D printing in water treatment. This is unachievable with traditional membrane preparation methods such as solvent casting, electrospinning, and phase inversion [5,7].

8.1.2 3D printed membrane for oil separation

Oil spillage during oil exploitation, extraction, and transportation is one of the serious causes of water pollution. The performance of traditional treatment methods such as physical adsorption, in-situ burning of oils, coagulation of chemicals, and dispersant treatment is often improper for the effective removal of oils from water, and in addition, all these are very expensive methods. The membrane-based separation method is potent in this sense, due to its exceptional wetting property such as super-

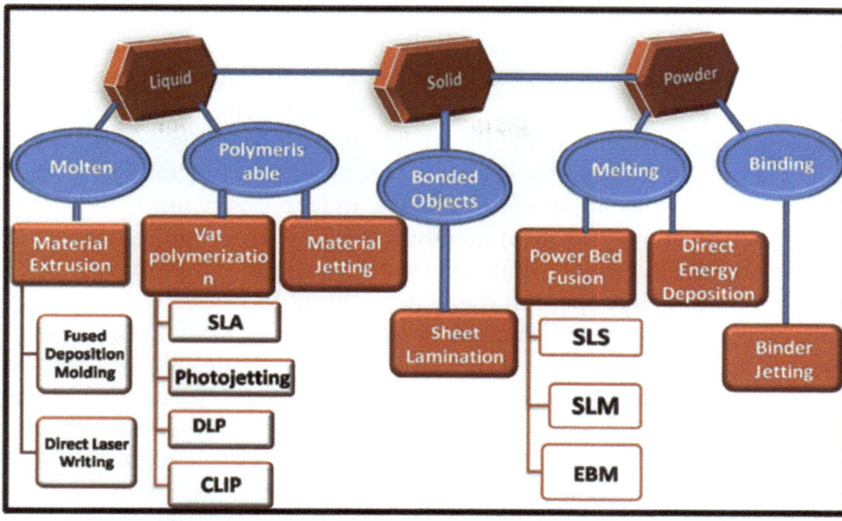

FIGURE 8.1 Different types in 3D Printing techniques. *Reproduced by permission from Springer Nature, [Environ. Sci. Pollut. Res.], [Review of manufacturing three-dimensional-printed membranes for water treatment, Issac, M.N., Kandasubramanian, B., 27, 36091–36108], Copyright (2020).*

hydrophobicity and super-oleophilicity [8]. These membranes possess high separation efficiency, high flux, and recyclability. The highly porous nature of the membranes provides efficacious removal of oil from water due to the large surface-to-volume ratio. However, the superhydrophobic membranes are prepared via a complicated multi-step method and are a tedious process. Usually, it involves the coating of micro/nano roughness layer on substrates such as mesh or fabric by spray coating, electrodeposition, surface etching, and solution mixing, followed by chemical modification of these substrates with low surface energy compounds, finally producing a superhydrophobic membrane. The methods used for coating the micro/nano components may sometimes lead to the agglomeration of these components on the mesh surface. This process may bring some sort of drawbacks in the internal properties as well as super-hydrophobicity of the membranes either due to the exfoliation and abrasion caused by limiting the adhesion between substrate and coating layer or by low cohesive force applied while coating the layer. As a whole overall performance of the membrane was restricted. Moreover, lack of reliability, limited selectivity, and expense are other disadvantages. The development of 3D printing technology helps to overcome limitations to a great extent by tuning the morphological and superhydrophobic properties of membranes manually with the help of a computer program. In addition, membranes with the required properties, including superhydrophobic properties, can be designed as required and the output delivered in a single-step process. There are polymers, ceramics, and alloys used as materials for the preparation of 3D printed membranes. Among 3D printed membranes based on polymers and ceramics are frequently used in oil/water separation. For the effective separation of oil from water, the membrane should possess surface wettability (super-hydrophobicity and super-oleophilicity), high porosity, and roughness. As mentioned above, the ideal membrane with a super-hydrophobic and super-oleophilic nature facilitates the easy penetration of oil into the pores of the membrane by repelling the water content [8].

Lv et al. developed a mechanically stable ordered mesh structure with nano-silica/polydimethylsiloxane (NS/PDMS) using the inkjet 3D printing method for the separation of oils from water [9]. The membranes are prepared by varying the concentration of silica, and the resultant NS/PDMS membrane with 18% and 20% silica possesses a highly ordered porous structure and is superhydrophobic in nature, without any collapse of silica particles. Furthermore, the superhydrophobic property of the membrane withstands mechanical stretching and bending (Fig. 8.2A). The 3D printing method facilitated the creation of roughness in the submillimeter scale by adjusting the spacing between the filaments, achieving super-hydrophobicity.

In addition, synthesized NS/PDMS membrane possesses super-oleophilic nature. This is due to the controlled pore sizes of the 3D printed membrane, which facilitates the efficient removal of hexane oil (Fig. 8.2B). It is found that within 30 ms, the oil is wet into the membrane pores. The pore size of the membrane is tailored readily by a computer program. This makes the oil-water separation more feasible. Furthermore, the water contact angle of the 3D printed membrane was found to be increased with the increase in pore size, leading to the increased water repellence rate. The capability of the membrane to separate different oils is depicted in Fig. 8.2C. Thus, the separation efficiency of the printed membrane is about 99.6% with a pore size of 0.37 nm and the flux is 23,700 $L\ m^{-2}\ h^{-1}$.

Another study was reported by Li et al. using the 3D inkjet printing method (Fig.8.3) [10]. In this study, they prepared a solid-like solution ink for printing, which is a mixture of cellulose acetate (CA), poly (vinyl alcohol) (PVA), and nano-silica particles (SiO_2 NPs).

The low water contact angle on air (18.14°) and high under water-oil contact angle pointed to the excellent super-hydrophobic and super-oleophilic nature of the 3D printed membrane. It also showed 99% oil/water separation efficiency after 50 cycles. Meanwhile, the composite membrane exhibits high mechanical stability accompanied by antifouling characteristics compared to those CA/PVA/Si meshes prepared via conventional methods.

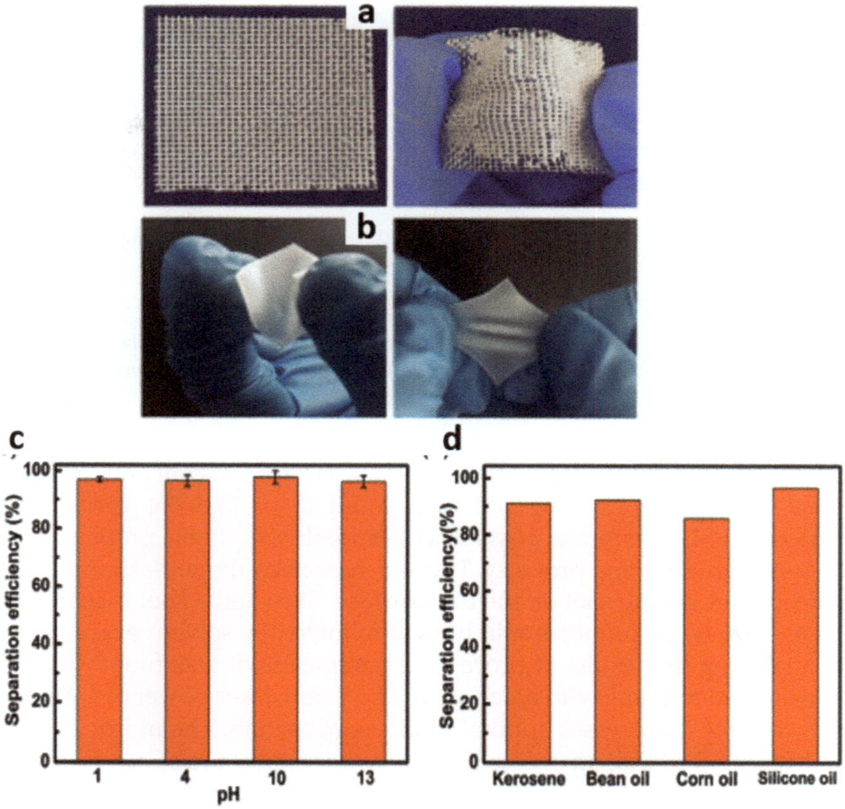

FIGURE 8.2 Images of (A) the mesh with the superhydrophobic coating before and after bending and (B) the printed membrane under bending and stretching. (C) Separation efficiency of the printed superhydrophobic membrane (with a pore size of 0.37 mm) for the mixtures (1:1, v/v) of hexane and nonneutral aqueous solutions with different pH values. The pH values of the aqueous solutions were adjusted by adding HCl or NaOH solution. (D) The separation efficiency of the superhydrophobic membrane for mixtures (1:1, v/v) of water and high viscosity oils, including kerosene, bean oil, corn oil, and silicone oil. *Reproduced by permission from Royal Chemical Society, [J. Mater. Chem. A], [3D printing of a mechanically durable superhydrophobic porous membrane for oil—water separation, Gong, Z., He, Z., Yang, J., Chen, Y., Tang, C., Liu, Y., Fan, M., Lau, W.-M., 5, 12435—12444], Copyright (2017).*

Polysulfone-based membranes are often used in water treatment for oil separation due to their excellent resistance to chemicals and high mechanical stability. Yuan et al. fabricated polysulfone (PSU) membranes via selective laser sintering 3D printing technique [11]. The authors utilized polysulfone with high amorphous content because it withstands a high temperature associated with high chemical resistance. The successful development of PSU membrane through laser sintering always depends on the processing parameters such as laser power, scan count, and hatch

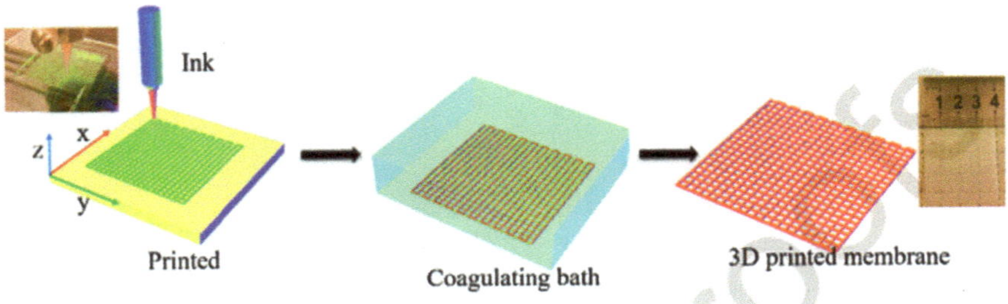

FIGURE 8.3 The inkjet 3D printing of composite membrane. *Reprinted from Separation and Purification Technology, 237, Li, X., Shan, H., Zhang, W., Li, B, 3D printed robust superhydrophilic and underwater superoleophobic composite membrane for high efficient oil/water separation, 116324; Copyright (2020) with permission from Elsevier.*

space. In this study, the laser power (15W) was kept constant and the hatch space (0.8 mm or 1.5 mm) and scan count (1 or 2) were varied. The fabricated PSU membrane possesses the superhydrophobic property with the switchable wetting property. This is achieved by the self-assembling of hydrophobic candle soot on the PSU surface. The candle soot is coated on the membrane by immersion in hexane followed by solvent evaporation. By analyzing the results of processing parameters, it was found that the fabricated membrane with hatch space of 1.5 mm, laser power of 15W, and scan count of 1 possesses optimal candle soot coating. The morphological analysis showed that an ordered loose network structure of candle soot is formed over the PSU membrane. The candle soot-coated printed PSU membrane exhibits a super-oleophilic character also. The water contact angle on the top surface of the 3D printed PSU membrane is increased after coating with candle soot from 89° to 160° (Fig. 8.4A and B). The separation efficiency of candle soot-coated printed PSU membrane is about 99% after 10 cycles for all hexane/water mixtures. The intrusion pressure of oils is 3.2 KPa for the membrane prewetted with water and it is around 2.9 KPa for candle soot-coated membrane prewetted with oils [11].

Biocompatible and biodegradable polymers have always been promising candidates for water treatment. Among these, poly(lactic acid) (PLA) is clinically approved by Food and Drug Administration (FDA) and is a water-soluble polymer having hydroxyl and carboxyl groups at the terminals. So, the preparation of the superhydrophobic membrane with PLA is always a challenging task. However, many researchers made attempts to prepare PLA membranes using conventional methods such as layer-by-layer coating, plasma treatment, and phase inversion. With the advent of 3D printing technology, new modifications were brought into the preparation of PLA membranes. Xing et al. synthesized porous PLA membrane for oil separation through the 3D printing technique [12]. The

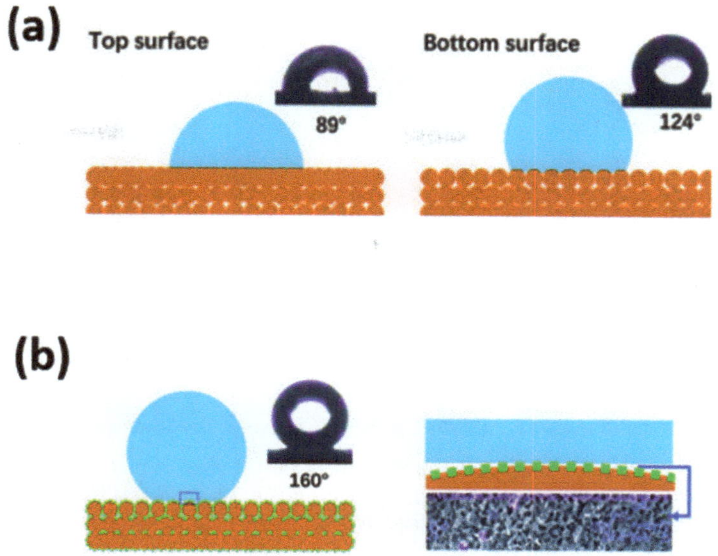

FIGURE 8.4 *Under process* Scheme of (A) water contact angle on the bottom and top surface (B) water contact angle on the candle soot-coated PSU membrane. *Reproduced by permission from Royal Chemical Society, [Journal of Material Chemistry A], [Super-hydrophobic 3D printed polysulfone membranes with a switchable wettability by self-assembled candle soot for efficient gravity-driven oil/water separation, Yuan, S., Strobbe, D., Kruth, J.-P., Van Puyvelde, P., Van der Bruggen, B., 5, 25401–25409], Copyright (2017).*

morphological structure of the membrane was designed on inspiration from the lotus surface structure and is implemented on the membrane by 3D printing followed by facile two-step chemical etching. The etchants used are acetone, ethyl acetate, DMF, and DCM. The as-synthesized membrane contains a hierarchical structure with nano-micro scale pores with a rough texture. Depending on the type of the etchants, the topographical structure and pore size of the 3D printed PLA membrane can be tailored considerably as shown in Fig. 8.5.

The oil separation efficiency of the 3D printed superhydrophobic PLA membrane is 99% for the pore size of 250 μm accompanied by a flux of 60 kL/m^2/h. The super-hydrophobicity of the membrane for different solutions is shown in Fig. 8.6. The study also found that the PLA membrane exhibits high chemical resistance, mechanical stability, and cost-effectiveness compared to traditional methods.

Ceramics are another class of material widely used in the separation of oil from water [13]. The traditional methods for the preparation of ceramic-based membranes suffer from improper alignment of transport channels in both separating layers and support, and poor adhesion between them. The implementation of the 3D printing technique paved a unique pathway for the fabrication of ceramic membranes with attractive

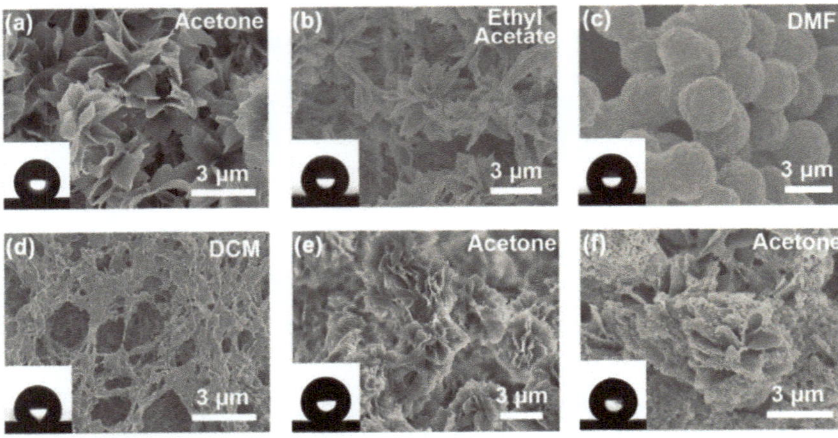

FIGURE 8.5 Surface morphology of 3D-printed PLA surface etched with various etchants. *Reproduced by permission from Xing, R., Huang, R., Qi, W., Su, R. and He, Z., 2018. Three-dimensionally printed bioinspired superhydrophobic PLA membrane for oil-water separation.* AIChE Journal, *64, 3700–3708, John Wiley and Sons.*

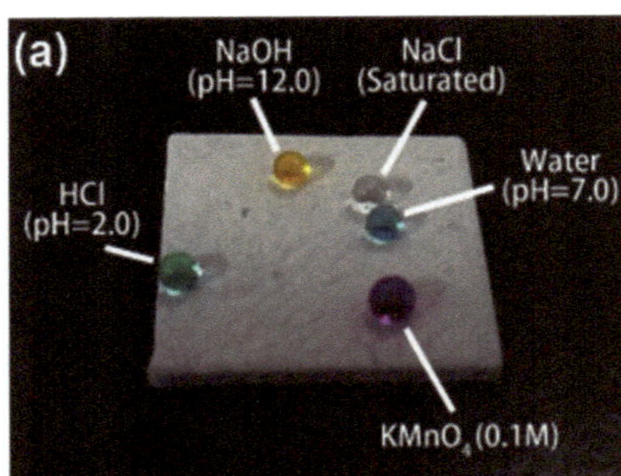

FIGURE 8.6 Superhydrophobic property, liquid drops on PLA surface. *Reproduced by permission from Xing, R., Huang, R., Qi, W., Su, R. and He, Z., 2018. Three-dimensionally printed bioinspired superhydrophobic PLA membrane for oil-water separation. AIChE Journal, 64, 3700–3708, John Wiley and Sons.*

properties and high uniformity [5]. Moreover, the incorporation of microporous materials enhanced roughness in the micro-meter size and thereby facilitated the excellent superhydrophobic character to the membrane. Whiskers-coated 3D printed ceramic membranes were reported by Chen et al. [13]. The membrane preparation and the application of fabricated membrane for the oil/water separation, oil purification, and oil/water emulsion separation are shown in Fig. 8.7.

Aluminum borate whiskers ($Al_{18}B_4O_{33}$) coated on printed alumina structures provided super hydrophilicity as well as superoleophobicity underwater to the ceramic membrane. As a result, the ceramic membrane showed excellent resistant chemicals, guaranteed robust, and efficient

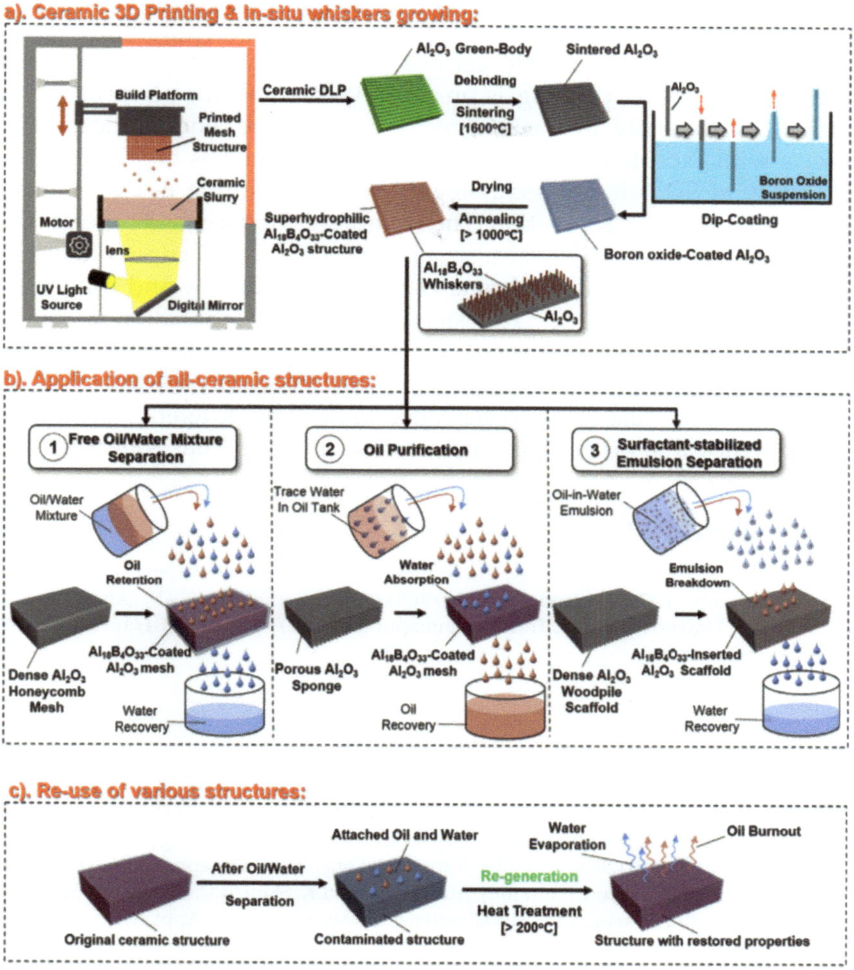

FIGURE 8.7 Illustration for the whiskers-coated ceramic mesh fabrication process and oil/water separation strategies. (A) Ceramics digital light processing and work flow of whiskers coating. (B) three kinds of printed ceramic structures for separation of free oil/water mixture, oil purification, and emulsion, respectively. (C) restoring the properties of the membrane via heat treatment. *Reprinted from Chemical Engineering Journal 373, Chen, Z., Zhang, D., Peng, E., Ding, J., 3D-printed ceramic structures with in situ grown whiskers for effective oil/water separation. 1223–1232; Copyright (2019) with permission from Elsevier.*

filtration performance. In addition, the separation efficiency of the membrane can be restored by facile heat treatment and is shown in Fig. 8.7C. The tailored structure of the printed whiskers coated ceramic membrane showed a separation efficiency of 99% and 99.7% for oil/water mixture and oil/water emulsion, respectively. They have remarkable resistance toward harsh environments and exceptional mechanical stability.

8.1.3 Conclusion and future perspectives

Oil spillage is one of the serious causes of water pollution. The development of 3D printing is an effective method for oil clean-up in water. The diversity in the designing, architecture, and fabrication of objects with exceptional properties and complex shapes highlighted the feasibility of 3D printing technology. Now, 3D printing is a prominent method in oil/water separation. The membrane prepared via different 3D printing techniques possesses a super-hydrophobic and super-oleophilic nature. The 3D printed membrane exhibits high porosity and roughness on the micrometer scale. The stepwise layered fabrication methods brought unique and uniform properties to the membrane. In brief, as an innovative technique in the field of water treatment, only a few studies are reported yet. So, vast research is needed to develop different membranes for the removal of other toxic contaminants and oils from water.

Acknowledgments

We are grateful to the APJ Abdul Kalam Technological University (KTU), Thiruvananthapuram (KTU proceedings No.4/1654/2019), and CERD, APJ Abdul Kalam Technological University (KTU), Thiruvananthapuram (KTU/RESEARCH 4/1694/2021) for the financial assistance.

References

[1] S.P. Dharupaneedi, S.K. Nataraj, M. Nadagouda, K.R. Reddy, S.S. Shukla, T.M. Aminabhavi, Membrane-based separation of potential emerging pollutants, Separ. Purif. Technol. 210 (2019) 850–866, https://doi.org/10.1016/j.seppur.2018.09.003.

[2] L.D. Tijing, J.R.C. Dizon, I. Ibrahim, A.R.N. Nisay, H.K. Shon, R.C. Advincula, 3D printing for membrane separation, desalination and water treatment, Appl. Mater. Today 18 (2020) 100486, https://doi.org/10.1016/j.apmt.2019.100486.

[3] N. Fijoł, A. Aguilar-Sánchez, A.P. Mathew, 3D-printable biopolymer-based materials for water treatment: a review, Chem. Eng. J. 430 (2022) 132964, https://doi.org/10.1016/j.cej.2021.132964.

[4] Z.-X. Low, Y.T. Chua, B.M. Ray, D. Mattia, I.S. Metcalfe, D.A. Patterson, Perspective on 3D printing of separation membranes and comparison to related unconventional fabrication techniques, J. Membr. Sci. 523 (2017) 596–613, https://doi.org/10.1016/j.memsci.2016.10.006.

[5] M.N. Issac, B. Kandasubramanian, Review of manufacturing three-dimensional-printed membranes for water treatment, Environ. Sci. Pollut. Res. 27 (2020) 36091–36108, https://doi.org/10.1007/s11356-020-09452-2.

[6] N. Shahrubudin, T.C. Lee, R. Ramlan, An overview on 3D printing technology: technological, materials, and applications, Procedia Manuf. 35 (2019) 1286–1296, https://doi.org/10.1016/j.promfg.2019.06.089.

[7] H. Dommati, S.S. Ray, J.-C. Wang, S.-S. Chen, A comprehensive review of recent developments in 3D printing technique for ceramic membrane fabrication for water purification, RSC Adv. 9 (2019) 16869–16883, https://doi.org/10.1039/C9RA00872A.

[8] U. Baig, M. Faizan, M. Sajid, Multifunctional membranes with super-wetting characteristics for oil-water separation and removal of hazardous environmental pollutants from water: a review, Adv. Colloid Interface Sci. 285 (2020) 102276, https://doi.org/10.1016/j.cis.2020.102276.

[9] J. Lv, Z. Gong, Z. He, J. Yang, Y. Chen, C. Tang, Y. Liu, M. Fan, W.-M. Lau, 3D printing of a mechanically durable superhydrophobic porous membrane for oil–water separation, J. Mater. Chem. 5 (2017) 12435–12444, https://doi.org/10.1039/C7TA02202F.

[10] X. Li, H. Shan, W. Zhang, B. Li, 3D printed robust superhydrophilic and underwater superoleophobic composite membrane for high efficient oil/water separation, Separ. Purif. Technol. 237 (2020) 116324, https://doi.org/10.1016/j.seppur.2019.116324.

[11] S. Yuan, D. Strobbe, J.-P. Kruth, P. Van Puyvelde, B. Van der Bruggen, Superhydrophobic 3D printed polysulfone membranes with a switchable wettability by self-assembled candle soot for efficient gravity-driven oil/water separation, J. Mater. Chem. 5 (2017) 25401–25409, https://doi.org/10.1039/C7TA08836A.

[12] R. Xing, R. Huang, W. Qi, R. Su, Z. He, Three-dimensionally printed bioinspired superhydrophobic PLA membrane for oil-water separation, AIChE J. 64 (2018) 3700–3708, https://doi.org/10.1002/aic.16347.

[13] Z. Chen, D. Zhang, E. Peng, J. Ding, 3D-printed ceramic structures with in situ grown whiskers for effective oil/water separation, Chem. Eng. J. 373 (2019) 1223–1232, https://doi.org/10.1016/j.cej.2019.05.150.

CHAPTER

9

Use of 3D printed techniques for organic pollutants removal

Ashwani Kumar[1], *Sunil Dutt*[2], *Raj Kumar*[3], *and Sonika*[4]

[1] Department of Chemistry, Govt. College Kullu, Himachal Pradesh, India; [2] Department of Chemistry, Govt. Post Graduate College Una, Himachal Pradesh, India; [3] Department of Pharmaceutical Sciences, University of Michigan, Ann Arbor, MI, United States; [4] Department of Physics, Rajiv Gandhi University, Itanagar, Arunachal Pradesh, India

9.1 Introduction

The advancement of science and technology in recent years has led to change in our lifestyle. But, worldwide human activities, along with advancements in farming, irrational consumption, advanced industrialization in different fields, population explosion, has led to environmental pollution, specifically organic waste pollution. Organic waste pollution arises from different sources like dye industry, pharma industry, and so on, which leads to major issues such as carcinogenicity. Non-biodegradability and toxicity of pollutants are major concerns [1,2], which damage the planet irreversibly. So, in modern era, organic pollution is not just a major concern but also leads to different major problems and diseases for public health, like neurological and endocrine disorder, reproductive and hormonal disorder, or various types of cancer, diarrhea. Moreover, organic pollutants have a negative impact on the global ecosystem causing illness and death worldwide [3,4]. Thus they are extremely hazardous to human health and ecosystem also [5–7]. They have greater negative environmental impact; so, the removal of organic pollutants is very important in environmental protection measurements. Several methods such as activated carbon adsorption, photocatalysis,

3D Printing Technology for Water Treatment Applications
https://doi.org/10.1016/B978-0-323-99861-1.00006-0

chemical oxidation, biological treatment, catalytic wet oxidation, biosorption, functionalized carbon nanomaterials were developed for organic pollutants removal, but the conventional methods of treatment find different difficulties [8,9]. In 1977, the 3D printing concept was started when a patent was filed and granted in name of W. K. Swainson for "using a laser to create covalent cross-linking at the surface of a liquid monomer where the object being manufactured rested on a tray that was gradually lowered into a vat one step at a time" [10]. Three-dimensional printing assisted with laser's liquid photopolymer solutions was started by H. Kodama of Nagoya Municipal Research Institute in 1981. Charles Hull explored the 3D printing technique to produce smaller versions of products. Scott Crump developed the fused deposition modeling (FDM) that is also known as desktop 3D printing method which is one of the most common 3D printing techniques till date [11]. Hence, the search for efficient and cost-effective material for pollutant removal ended with 3D printing technique, also known as additive manufacturing technique. Moreover, disadvantages of already existing methods have driven for new technologies development and the 3D printing techniques inspired different research groups, which has attracted researchers and scientists during recent years. Remediation of wastewater using new types of 3D nanostructured materials was explored worldwide. These materials offer facile and cost-effective method with less waste production, low fabrication cost, rapidity for organic pollutants removal due to their high efficiency (>75%) and reusability (over 50 cycles) for metal-organic frameworks with decorated hierarchical porous ceramics, flexibility to modify surfaces of materials [12]. Three-dimensional objects fabrication was started with the emergence of this area and nowdays it is transversal to several areas [13]. In order to manufacture 3D material of desired shape, the material is deposited layer-by-layer using 3D printer by bottom-up method and the schematic diagram is shown in Fig. 9.1. Various types of 3D printers are available now days such as Laminated Object Manufacturing (LOM), stereolithography (SLA), photopolymer inkjet printing, Fused Deposition Modeling (FDM), digital light processing (DLP), etc. [14] (Fig. 9.1).

So, this concept is used for the manufacturing of different 3D objects. Three-dimensional hierarchical nanoporous Cu (3D NP-Cu) was synthesized using chemical dealloying 3D printing laser technique. Larger specific surface area with many folds than conventional catalysts around 660-fold offer very effective method for degradation of organic pollutants [15]. Aerogels based on graphene and biopolymer for water treatment were prepared using direct ink writing (DIW) as 3D printing technique

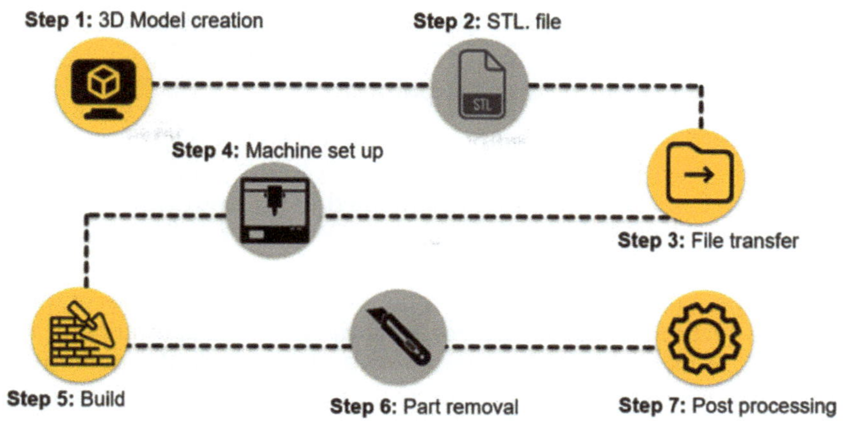

FIGURE 9.1 Schematic representation 3D printing of an object with desired shape. *Reproduced with permission from H. Agrawaal, J.E. Thompson, Additive manufacturing (3D printing) for analytical chemistry, Talanta Open 3 (2013) 100036. doi:10.1016/j.talo.2021.100036. Copyright 2021, Elsevier. Open access.*

and freeze-drying subsequently, and this study was published as proof of concept [16]. Hundred percent removal of organic solvents with better regeneration and reusability was reported using these aerogels.

Chemical dealloying assisted laser 3D printing technique was used to manufacture Cu three-dimensional (3D) hierarchical nanoporous structures with different geometries. The specific surface area of these Cu (3D NP-Cu) nanostructures was found to be many fold (660-fold) larger than conventional 2D ribbons with better degradation efficiency, good reusability, excellent versatility toward azo dyes with a kinetic reaction constant of 0.147 min^{-1} [15].

Low-density-polyethylene (LDPE) was decorated with TiO_2 to synthesize 3D-printed floating photocatalyst with less density than water and more stability [17]. Ceramic membrane-based treatment materials were also manufactured using 3D methodology, and they also have multiple hierarchy design [11,18]. Three-dimensionsal printer with PLA filament was used to prepare polylactic acid (PLA) sorbent modified with carbon black. This sorbent showed internal porosity from 15% to 50%, which was controlled through software. This sorbent showed better removal of VOC contaminants such as benzene, toluene, and ethyl benzene [19]. A self-powered electro-Fenton (EF) system integrated (3D printed elastic triboelectric nanogenerator (PE-TENG)) with the biomass-based carbon materials as cathode having good superhydrophilicity, high surface area, and large defect degree was used to remove methylene blue with efficiency of 97.0% within 140 min [20].

9.2 Organic pollutants: classification and adverse effects

9.2.1 Classification of organic pollutants

We know that organic compounds contain carbon-hydrogen bonds. Hence any chemical pollution caused by carbon pollutants is called *organic pollution*. The pollutants are defined as harmful contents which directly affect the environment and through this affect the human life and living organism [21].

Depending on their condition, a pollutant can quantitatively or qualitatively affect the environment. For instance, if the concentration of carbon dioxide increases in the environment, we can say the quantity of carbon dioxide is increased and it becomes harmful to human life [22]. On the other hand, e.g., insecticide, a pollutant which is not present in the environment but added by human activities, can affect human life and living organism qualitatively [23].

The organic pollutants include chlorinated phenols, phenols, azo dyes, endocrine disrupting chemicals, polychlorinated biphenyls, polyaromatic hydrocarbons, pesticides, etc. [24].

9.2.1.1 Organic pollutants can be further classified into following categories

9.2.1.1.1 Quantitative and qualitative pollutants

9.2.1.1.1.1 Quantitative pollutants These types of pollutants are categorized on the basis of their quantity. They occur in nature and become pollutants when their concentration reaches beyond a certain value. It means their quantity in nature has increased, e.g., carbon dioxide, oxides of nitrogen, etc. As all of us know, carbon dioxide is essential for plants and required in photosynthesis process [25]. But, when the concentration of carbon dioxide in the nature has increased, it increases the Earth's temperature and leads to global warming [26].

9.2.1.1.1.2 Qualitative pollutants These are man-made organic chemicals and do not occur in nature. These chemicals enter the environment through human activities. These are toxic in nature. The examples of these qualitative pollutants are insecticide, herbicides, fungicides, DDT, etc.

9.2.1.1.2 Primary pollutants and secondary pollutants

9.2.1.1.2.1 Primary pollutants These pollutants which are directly released from source into atmosphere or from human activities on the Earth are called primary pollutants. They remain in unmodified forms. Burning of fossil fuels causes the release of carbon monoxide (CO), hydrocarbons, nitrogen oxides (NO_x), sulfur dioxide (SO_2), and particulate matter. For example, burning of organic material such as coal, gas, wood, etc., and exhaust from automobiles are the biggest sources the CO. CO is toxic because it reduces oxygen in blood due to its combination with hemoglobin, but not regarded as persistent organic pollutant, because it forms carbon dioxide with oxygen [27]. Similarly, incomplete burning of hydrocarbons causes air pollution. These hydrocarbons also contribute to smog. The major source of hydrocarbon is automobiles and some industries. Moreover, particulates which are liquid droplets (~2.5–10 mm) and small pieces of solid materials cause pollution; examples include asbestos from brakes, ash from fires, and insulation [28]. Ability of lungs to exchange gases gets hindered due to their accumulation in lungs.

9.2.1.1.2.2 Secondary pollutants Air pollutants which are formed in atmosphere as a result of chemical or physical reaction with primary pollutants or other atmospheric components are known as secondary pollutants. Examples include photochemical oxidants like PAN (peroxy acetyl nitrate), ozone, aerosols, and photochemical smog, mists, etc. [29]. The chemical products of secondary pollutants are highly reactive and can cause severe effects on living beings and ecosystem.

9.2.1.1.3 Volatile pollutants and semivolatile organic pollutants

The volatile pollutants and semivolatile organic pollutants are mostly synthetic organic compounds and have been originated due to lack of their management and also from industrial activities. Boiling points can be one parameter for the classification of volatile and semivolatile organic compounds. The classification reveals the differences in the chemical properties and methods for their quantitative and qualitative detection.

9.2.1.1.3.1 Volatile organic compounds Organic compounds such as methane, propane, butane, benzene, xylene having low boiling points are under this class and hence can be evaporated under normal conditions of temperature and pressure. Therefore, they can enter into the atmosphere easily. Petrochemicals industries, agricultural source (mainly methane derivatives), organic solvents, etc., are the major contributors of VOCs [30].

9.2.1.1.3.2 Semi-volatile organic compounds This category includes organic compounds having boiling points higher than volatile range and up to 400°C. The examples of semi-volatile organic pollutants include a myriad of industrial chemicals such as petrochemicals, pesticides, industrial chemicals and carbonyl compounds, phenols, ethers, aromatic and aliphatic esters, pyridines, anilines, and many others.

9.2.1.1.4 Nonpersistent and persistent organic pollutants

A pollutant that can be removed, decomposed, or consumed and thus reduced to tolerable level either by engineered techniques or natural processes. These pollutants are of two types: rapidly degradable (nonpersistent) and slowly degradable (persistent).

9.2.1.1.4.1 Nonpersistent pollutants Rapidly degradable pollutants can be removed rather quickly if the system is not overloaded. These can be broken down by natural cycling processes. The examples of persistent pollutants include sewage plants, synthetic organic compounds, some metals and some metal salts, etc.

9.2.1.1.4.2 Persistent organic pollutants (POPs) POPs are slowly degradable organic chemical material take long time to decompose, sometimes also called "forever chemicals." POPs are unaffected by environmental degradation, over any chemical, physical, or biological change. POPs bioaccumulate in environment and have harmful effects on human health and wildlife [31]. The Stockholm Convention on Persistent Organic Pollutants in 2001 discussed the effects and consequences of POPs on human and environmental health. The intention of this convention was to eliminate or severely restrict the production of POPs, by the international community. Examples of POPs include waste water (oxygen demanding waste), DDT, pesticides, radioactive substances and synthetic organic compounds, hydrocarbons, HCBs, PCBs, Mirex, Chlordane, endrin, etc.

Some of the examples of POPs are discussed below:

Oxygen demanding wastes In general, oxygen demanding wastes are water pollutants that are decomposed by oxygen-requiring bacteria. In most cases, oxygen demanding wastes generally come from wastewater. The waste is a byproduct of domestic, commercial, agricultural, and industrial activities. The wastewater can be categorized depending upon its source. It includes domestic wastewater from households, municipal sewage from the communities, wastewater from food industries, animal feedlots, pulp mills, and run-off from farmlands as well as dumping of wastes into streams. Besides this, wastes from dead plant and animal tissues also add immensely to oxygen demanding wastes. These wastes consume oxygen dissolved in water during the course of degradation. These oxygen demanding wastes decrease the level of dissolved oxygen (DO); this is an indication of pollution due to organic matter, of above said examples [32]. All these materials undergo degradation by microbial activities in presence of DO. Hence, reduction of the DO is a serious problem which adversely affects aquatic life, if its concentration falls below 4.0 mg/L. This reduction of DO is an index of pollution as well.

Synthetic organic compounds Carbon-based manmade compounds are known as synthetic organic compounds (SOCs). Various SOCs are produced in the laboratories. These compounds are used globally in huge quantities for diverse purposes, and therefore, they are frequently released into the environment. Furthermore, SOCs also enter into the ecosystem through several manmade activities which includes manufacture of these compounds, their leakage during the transportation, and their uses in variety of applications.

Synthetic (manmade) organic chemicals are either intentionally or nonintentionally produced/released in the environment. These chemicals remain in environment and it takes hundreds of years for their degradation. These SOCs include synthetic pesticides, synthetic detergents, synthetic fibers, pharmaceuticals, food additives, plastics, solvents, and volatile organic compounds. Few examples of pesticides are aldrin, dieldrin, chlordane, DDT, endrin, heptachlor, mirex, toxaphene, etc., and industrial chemicals such as hexachlorobenzenes (HCBs) and polychlorinated biphenyls (PCBs) [33].

Some of the examples of these SOCs are listed in Fig. 9.2. These pesticides remain in environment and cause pollution because they are resistant to microbial degradation. Even traces of these pesticides make water unfit for use. PCBs, which have been used since 1930s are complex mixtures of chlorobiphenyls, can degrade or breakdown in the environment but the process of their breakdown depends on the chemical structure of the PCBs. Usually, PCBs are broken down by sunlight or by particular microorganisms in the environment. Because of high lipophilicity of the PCBs, they travel through the environment as well as within the tissues or cells. Once introduced into environment, these compounds are exceptionally persistent and are very highly stable to chemical reagents.

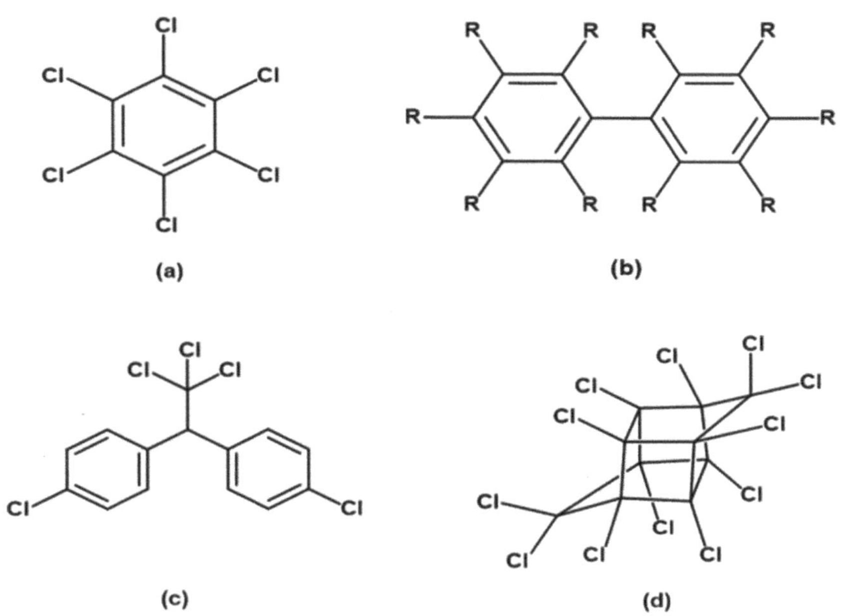

FIGURE 9.2 Some common examples of organic pollutants. (A) HCB (B) PCB (C) DDT and (D) Mirex.

9.2.1.2 *Adverse effect of organic pollutants*

Organic pollutants affect human life, aquatic life, wildlife, and environment adversely. Although we have already discussed the adverse effects of organic pollutants, some of the major effects of these pollutants are listed below:

1. The organic pollutants bond strongly with fat tissues of fish and other aquatic organisms, milk of mammals, etc.
2. These pollutants contaminate air, soil, and water, and enter through food chain and affect people's health adversely. People are generally exposed to POPs through polluted foods.
3. Among people, developmental, neurological, behavioral, and immunological adverse health effects have been linked to POPs.
4. Exposure to POPs during pregnancy has been linked to development effects in offspring.
5. The pollutants persist in the system for decades causing problems such as cancer, birth defects, immunological and reproductive disorder in human and animals.

9.3 Techniques to detect organic pollutants and their removal

Organic pollutants in the ecosystem, especially POPs, leads to many environmental problems in the world. In the past few decades, there has been a high increase in production and utilization of organic pollutants resulting in big threat of pollution. Several efficient techniques have been developed for the removal of highly toxic organic pollutants from the environment. Some of these techniques are listed below:

9.3.1 Separating components of a mixture by extraction or chemical method

This is also called chemical method of separation of organic pollutants. In this method, the pollutants can be removed with help of solvent polarity. The solvents with hydrophobic and hydrophilic nature are used in this technique. These solvents are having different polarities; one having high polarity like water and other have low polarity like hexane or ethyl acetate. Sometime surfactants can be used, which possess a hydrophobic head and hydrophilic tail. The organic pollutants if present are highly soluble in low polarity solvents and can be extracted from polar compounds. In this way, the pollutants can be removed easily. This is quite an inexpensive method for the removal of organic pollutants [34].

9.3.2 Test for the functional group

Organic pollutants can be removed by detecting the functional group too. The various tests can be performed for the detection of functional groups in laboratory. It is a traditional way of detecting the organic pollutants.

9.3.3 Quantitative and qualitative analysis

Another way of detecting organic pollutants is quantitative and qualitative analysis. In quantitative analysis we can easily quantify the pollutants, using volumetric analysis. On the other hand in qualitative analysis, we cannot quantify the organic pollutant but easily we can detect the pollutants using qualitative analysis, where we can do physical examination, detection of elements and functional groups present in the pollutants. This is also traditional way of detecting the organic pollutants.

9.3.4 Spectroscopic techniques

This is the modern method for the detection of organic pollutants. The spectroscopic technique includes UV-vis, infrared, nuclear magnetic resonance, and ESI-MS spectroscopy. The UV-vis is used to detect the presence of conjugation, chromophore, and auxochrome in the organic pollutants. On the other hand infrared spectroscopy is used to detect the presence of various functional groups. NMR is used to detect the nature and environment of the protons, kind of protons, carbons, and heteroatoms present in the pollutants. Nowadays, it is also possible to calculate the mass of pollutants with the help of ESI-MS techniques. Using UV-vis, IR, NMR, and ESI-MS techniques the complete structure of the pollutants can be detected easily [35].

After the detection of these pollutants, they can be removed by extraction method, or by column chromatography. However besides these tradition methods a new 3D printed techniques for organic pollutants removal can be used which is more effective in short time as compared to other traditional techniques. This technique is discussed in detail in the next section.

9.4 3D printed techniques for organic pollutant detection and removal

9.4.1 Detection

Environmental pollution is growing at an alarming speed, and causes irreversible damage to the planet. Change in lifestyle of the human beings in modern era has increased the environment pollution. The major problems we are facing worldwide are the environmental challenges such as contamination of ground water, hazardous waste coming from industries, etc. [36–38]. The numerous pollutants responsible for these problems are cosmetics, pesticides, steroid hormones, pharmaceuticals, and personal healthcare products [39]. Among the various environmental issues, organic pollutant is the major issue due to their adverse side effect and long-term impact on health. Among the various pollutants, organic pollutants are more dangerous to health due to their highly toxic nature and may induce adverse side effects including neurological, developmental, and immune effects [40]. The major organic pollutants are hydroquinone, malathion and acephate phenyl hydrazine, polychlorinated biphenyl, 17b-estradiol ethinylestradiol, cocaine, atrazine, and fenoxycarb [41]. Their toxicity and side effects in human are still not completely understood, though they are detected in human blood, fluids, adipose tissue, and breast milk. Hence, detection of pollutant is key strategy to manage the corresponding side effects. Various strategies have been developed to detect these pollutants. Conventional techniques have their own limitations along with their advantages as well as disadvantages such as time-consuming, limitation in detection, and suitable to detect only for limited organic pollutant.

Hence, research and development open a new window to develop the technique to solve it. In the past decade, several strategies have been developed. Numerous sensing techniques have been developed to detect the organic pollutants such as electrochemical, surface enhanced Raman spectroscopy, fluorescent, and gel electrophoresis [42]. One of the promising strategies to detect organic pollutants is by using 3D printing. Very recently 3D printing techniques have been developed and successfully applied for detection of various types of pollutants including organic pollutants.

Three-dimensional printing technology fabricates the designed part using CAD software that develops the object from the bottom to upper region. The computer design is translated into physical models by 3D printing through layered patterning of materials extruded via print head and nozzle [43]. Hideo Kodama for the first time introduced 3D printing through additive process in which solid object was created by hardening polymer using ultraviolet light [44]. Further, 3D printing is also called as additive processes, rapid prototyping, layer fabrication, additive fabrication, and solid free form fabrication. This 3D printing is used in various applications like manufacturing, architecture, and medicine, as well as pharmaceutical sector. Moreover, 3D printing has different types such as powder bed fusion, vat photopolymerization, direct energy deposition, material jetting, binder jet printing, and fused filament fabrication [45].

Among the various organic pollutants, United States Environmental Protection Agency (EPA) and the European Union (EU) considered phenol as key pollutant. Phenolic compounds are produced in pharmaceuticals, e.g., paracetamol produced p-aminophenol on hydrolytic degradation. Phenols also used for manufacturing dyes, rubber, petroleum, and photography. Moreover, it may cause serious threat. Hence, detection of phenol is more crucial. Cheng et al. developed the phenol detection technique using 3D printing; where 3D printed gold-plated electrode are used for simultaneous detection of phenol and p-AP in aqueous medium [46]. Balram et al. fabricated the 3D flower like zinc oxide decorated fMWCNTs through ultrasound for detection of 4-nitrophenol detection. Their results showed outstanding sensitivity, very low detection limit, and broad linear range. Real sample analysis is used to evaluate the feasibility of sensor. Fig. 9.3 represents the schematic of synthesis and detection of pollutants [47].

Due to ceramic and glass industries, uranium is a potential pollutant to water and soil. Its toxicity is associated with biomolecules such as peptide and proteins. The most common way for such pollutants to enter into the human body is inhalation of dust particles, water and food chain. It damages the renal and pulmonary systems and accumulates in the liver.

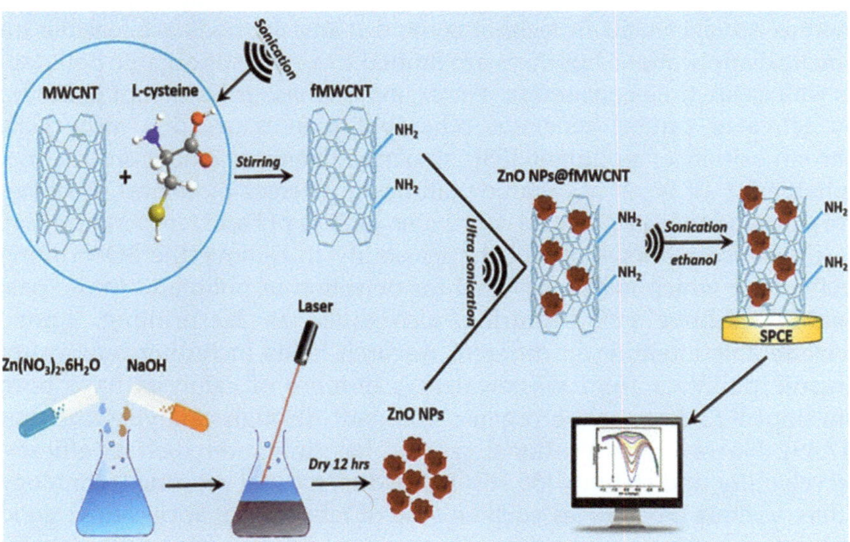

FIGURE 9.3 Schematic representation of Zno@fMWCNTs/SPCE fabrication using sonochemical process and its use in 4-NP detection. *Reproduced with permission from D. Balram, K.Y. Lian, N. Sebastian Ultrasound-assisted synthesis of 3D flower-like zinc oxide decorated fMWCNTs for sensitive detection of toxic environmental pollutant 4-nitrophenol, Ultrason. Sonochem. 60 (2020) 104798. doi:10.1016/j.ultsonch.2019.104798. Copyright 2019, open access, Elsevier.*

Urbanova et al. fabricated 3D-printed titanium electrode for sensing of uranyl ions in different pollutant samples in the range of 0–1000 μM [48]. All these results encourage further development of electrochemical sensing application of 3D printing technology. Nasir et al. fabricated 3D printed graphene electrodes for sensing the mycotoxins in food [49,50]. Kumar et al. fabricated 3D printed nanocarbon electrodes which were coated with Cu and Ni through electrochemical process (Fig. 9.3) which showed excellent sugars sensing performance [51]. Several organic pollutants detection techniques have been reported in literature majorly using 3D printing technology. Based on this 3D printing, we can detect the range of organic pollutants through various detection strategies. Long detection range, very low detection limit, and fast detection time are making the techniques more promising.

9.4.2 Removal

Detection of organic pollutant is key step toward the management of organic pollutant. Elimination or minimization of the adverse effects of pollutants can be achieved through its removal. However, removal of organic pollutant from wastewater is highly challenging. The most promising methods for removal of organic pollutants are activated carbon adsorption, chemical oxidation, and biological treatment [52]. However, there is advancement in technology and many methods are feasible for practical applications but these are limited to a range of organic pollutant as well as are time-consuming. For example, phase transition of pollutant by activated carbon generates other pollution issues due incomplete decomposition of pollutant [53]; similarly, chemical oxidations are not suitable for all types of organic pollutants, whereas biological processes are slow and required control conditions such as pH and temperature [54].

Then researchers explored the feasibility to remove the 3D printing technology which is already used for detection of pollutant. In the past decade, additive manufacturing, also called as 3D printing, gained considerable attention in different research fields including removal of organic pollutant from wastewater. A number of catalysts have been developed to remove the organic pollutant through catalytic reactions [17,49]. However, conventional catalyst has limitation such as efficacy, recyclability, and stability [55,56]. Three-dimensional printing technology offers various advantages such as ease of fabrication, application, good stability, excellent recyclability, efficacy, and high rate of reaction. Three-dimensional printing of foams, scaffolds, membranes have wide application in depuration of water where toxic pollutants penetrate the human body through the food chain [18,57,58]. Hence, may affect the function of major organs such as brain, heart, kidney, lungs, liver, and skin. This may induce cancers, allergies, dysfunction of organs. Among the scaffold and foams, membranes are the potential candidate due to semipermeable and poriferous properties. These properties allow to remove the organic pollutants such as dyes, oils, and pharmaceuticals [18,57,58]. Over the above techniques, 3D printing-based manufacturing of membranes has overcome the limitation of membranes manufacture through other techniques, such as pressure drop, concentration gradient, soiling, and low material transferring.

Liu et al. designed 3D printed hierarchical porous ceramics through in-situ loading of metal organic frameworks-based catalyst for hydrothermal treatment. The fabricated design was able to completely decompose the organic dyes through Fenton reaction with highest removal efficiency (99.68%) and rate of degradation (0.2915 min^{-1}). Due to large surface area, numerous active sites, hierarchical matter transport channels, and interconnected porous networks, it shows outstanding degradation performance toward organic pollutant such as dyes. Moreover, it is reusable up to 50 cycles with degradation efficiency >75% [12]. Masud et al. reported the direct ink-writing 3D printing technology for preparation of graphene biopolymer aerogels for treatment of water. They have added biopolymers polydopamine (PDA) and bovine serum albumin (BSA) to the bioink to form G-PDAPBSA aerogel. It showed superior water contaminant such as Cr(VI), Pd(II), cationic methylene blue, evans blue, n-hexane, and n-heptane, toluene in batch absorption studies. The aerogel showed 10 times recyclability with 100% removal of organic solvents and showed 2%–20% decrease after third cycle in removal of methylene blue. This study shows the potential of functional materials graphene efficacy as organic pollutant removal through 3D printing technology [16]. However, further studies need to be conducted to explore the efficacy of various functional nanomaterials. Shi et al. fabricated the film of PLA through 3D printing and coated with metal organic frameworks (Cu-MOFs) for removal of malachite green. It showed efficacy to remove malachite green absorbing at different concentration in 10 min. The film is recyclable more than 5 times through washing with acetone. This technique has an excellent potential for different application in organic pollutant removal and waste water treatment [59] (Fig. 9.4).

9.5 Advantages of 3D printed techniques

Development of manufacturing processes is making continuous progress. New techniques such as additive manufacturing, nanotechnology, and robotics are some examples of advancement in manufacturing techniques. Recently 3D printing technology made tremendous progress due to several advantages over other technologies. The major advantages of 3D

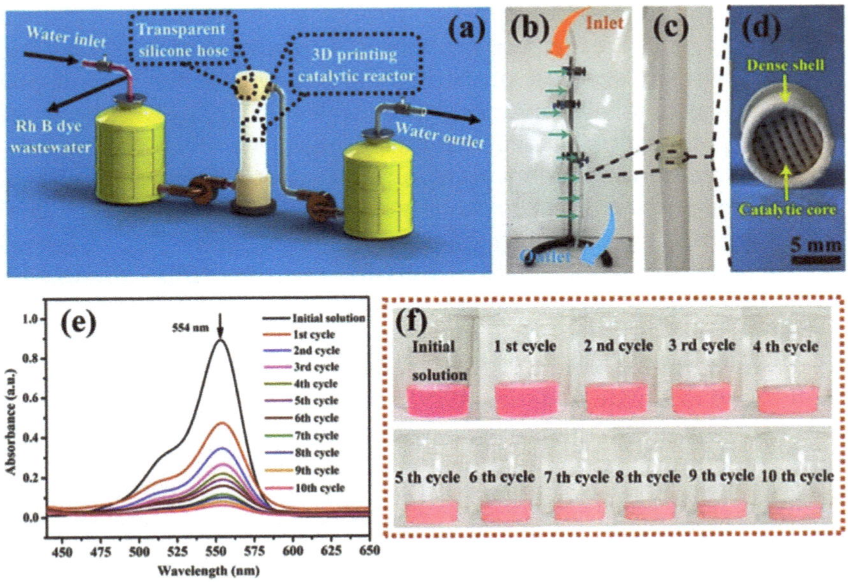

FIGURE 9.4 3D printed catalytic device for organic pollutant removal. (A) 3DP-HPC@MOFs degradation of dye from wastewater, (B–D) fabrication of catalytic device, (E–F) efficacy to remove Rh-B from wastewater and corresponding UV-vis spectrum change with cycle and color change of water. *Reproduced with permission from D. Liu, et al., 3D printing of metal-organic frameworks decorated hierarchical porous ceramics for high-efficiency catalytic degradation, Chem. Eng. J. 397 (2020) 125392. doi:10.1016/j.cej.2020.125392. Copyright 2020, Elsevier.*

printing are cost efficiency; ease of operation; design of complex structures; production in a single step; control over structures; flexibility; risk mitigation; sustainable, chemical, and mechanical stability; and minimal materials wastage. To produce prototypes and customized devices, industries widely use the 3D printing technology. Recently, the application of 3D printing expanded to diverse applications such as human organs fabrication, customized bone prosthetics, reaction ware for chemical synthesis, supercapacitors, and buildings construction. Due to development of affordable 3D printing techniques their use expanded to labs and even homes. This opens limitless opportunities for 3D printing technology applications.

9.6 Future perspectives of 3D printed techniques

Three-dimensional printing isn't a theoretical concept anymore and now it has become a small business in terms of manufacturing with implementation. Previously this area was not well explored and recently it has made revolution in different fields such as in aircraft industry, artificial human organs development, musical keyboards, and much more. In this technique, we can just imagine and create 3D model of desired object by clicking a button. For example, world's first entirely 3D-printed rocket was supposed to be launched into orbit by US-based aerospace startup. So, this leads to making a product with cost-effectiveness and superior quality. By 2025, 3D printing could generate economic impact of $230 billion to $550 billion per year across the globe as predicted by Wohler. So, this technology will change the manufacturing world. Due to usefulness in environmental applications, 3D printing is becoming very popular. This technique will find future applications in different areas such as wind turbines, filters, air quality monitors, membranes, microbial fuel cells, separation devices for water treatment, solar cells, etc. This book chapter may inspire the readers for the development of new types of 3D materials for remediation of organic pollutants.

9.7 Conclusion

Three-dimensional printing technology, also known additive manufacturing technique, has presented new opportunities for wastewater treatment. It has attracted considerable attention probably due their several advantages and potential benefits such as its simplicity and ability to rapidly produce analytical devices at low cost. The disadvantages of already existing conventional methods for multipollutant removal were overcome by the development of this technique. Thus, the benefits of these methods have attracted the attention of scientific community worldwide. This method will provide good opportunities for environmental remediation in future.

References

[1] J. Machhi, et al., Nanocarrier vaccines for SARS-CoV-2, Adv. Drug Deliv. Rev. (2021) 215–239, https://doi.org/10.1016/j.addr.2021.01.002. Elsevier B.V.

[2] Y. Qin, L. Zhang, T. An, Hydrothermal carbon-mediated fenton-like reaction mechanism in the degradation of alachlor: direct electron transfer from hydrothermal carbon to Fe(III), ACS Appl. Mater. Interfaces 9 (20) (2017) 17115–17124, https://doi.org/10.1021/acsami.7b03310.

[3] R. Kumar, et al., Core-shell nanostructures: perspectives towards drug delivery applications, J. Mater. Chem. B (2020) 8992–9027, https://doi.org/10.1039/d0tb01559h. Royal Society of Chemistry.

[4] R. Kumar, S.V. Dalvi, P.F. Siril, Nanoparticle-based drugs and formulations: current status and emerging applications, ACS Appl. Nano Mater. (2020) 4944–4961, https://doi.org/10.1021/acsanm.0c00606. American Chemical Society.

[5] S. Esplugas, et al., Ozonation and advanced oxidation technologies to remove endocrine disrupting chemicals (EDCs) and pharmaceuticals and personal care products

(PPCPs) in water effluents, J. Hazard Mater. (2007) 631–642, https://doi.org/10.1016/j.jhazmat.2007.07.073. Elsevier.

[6] J. Lintelmann, et al., Endocrine disruptors in the environment: (IUPAC technical report), Pure Appl. Chem. (2003) 631–681, https://doi.org/10.1351/pac200375050631. Walter de Gruyter GmbH.

[7] D. Pestana, et al., Persistent organic pollutant levels in human visceral and subcutaneous adipose tissue in obese individuals-Depot differences and dysmetabolism implications, Environ. Res. 133 (2014) 170–177, https://doi.org/10.1016/j.envres.2014.05.026.

[8] Z. Aksu, Application of biosorption for the removal of organic pollutants: a review, Process Biochem. (2005) 997–1026, https://doi.org/10.1016/j.procbio.2004.04.008. Elsevier.

[9] L.Y. Jun, et al., An overview of functionalised carbon nanomaterial for organic pollutant removal, J. Ind. Eng. Chem. 67 (2018) 175–186, https://doi.org/10.1016/j.jiec.2018.06.028.

[10] W.K. Swainson, Method, Medium and Apparatus for Producing Three-Dimensional Figure Product, 1971.

[11] H. Dommati, et al., A comprehensive review of recent developments in 3D printing technique for ceramic membrane fabrication for water purification, RSC Adv. (2019) 16869–16883, https://doi.org/10.1039/c9ra00872a.

[12] D. Liu, et al., 3D printing of metal-organic frameworks decorated hierarchical porous ceramics for high-efficiency catalytic degradation, Chem. Eng. J. 397 (2020) 125392, https://doi.org/10.1016/j.cej.2020.125392.

[13] C.G. Amorim, et al., 3D Printing Technology in the Environment, Springer, Singapore, 2021, pp. 131–160, https://doi.org/10.1007/978-981-15-8999-7_6.

[14] H. Agrawaal, J.E. Thompson, Additive manufacturing (3D printing) for analytical chemistry, Talanta Open 3 (2021) 100036, https://doi.org/10.1016/j.talo.2021.100036.

[15] C. Yang, C. Zhang, L. Liu, Excellent degradation performance of 3D hierarchical nanoporous structures of copper towards organic pollutants, J. Mater. Chem. 6 (42) (2018) 20992–21002, https://doi.org/10.1039/C8TA07973K.

[16] A. Masud, C. Zhou, N. Aich, Emerging investigator series: 3D printed graphene-biopolymer aerogels for water contaminant removal: a proof of concept, Environ. Sci. J. Integr. Environ. Res. Nano 8 (2) (2021) 399–414, https://doi.org/10.1039/d0en00953a.

[17] M.J. Martín de Vidales, et al., 3D printed floating photocatalysts for wastewater treatment, Catal. Today 328 (2019) 157–163, https://doi.org/10.1016/j.cattod.2019.01.074.

[18] L.D. Tijing, et al., 3D printing for membrane separation, desalination and water treatment, Appl. Mater. Today (2020) 100486, https://doi.org/10.1016/j.apmt.2019.100486. Elsevier Ltd.

[19] L.A. Lagalante, A.J. Lagalante, A.F. Lagalante, 3D printed solid-phase extraction sorbents for removal of volatile organic compounds from water, J. Water Proc. Eng. 35 (2020) 101194, https://doi.org/10.1016/j.jwpe.2020.101194.

[20] M. Tian, et al., Engineering flexible 3D printed triboelectric nanogenerator to self-power electro-Fenton degradation of pollutants, Nano Energy 74 (2020) 104908, https://doi.org/10.1016/j.nanoen.2020.104908.

[21] F.O. Ajibade, et al., Environmental pollution and their socioeconomic impacts, in: Microbe Mediated Remediation of Environmental Contaminants, Elsevier, 2021, pp. 321–354, https://doi.org/10.1016/b978-0-12-821199-1.00025-0.

[22] Managing Water Resources in the West Under Conditions of Climate Uncertainty, Managing Water Resources in the West under Conditions of Climate Uncertainty, National Academies Press, 1991, https://doi.org/10.17226/1911.

[23] M.A. Hassaan, A. El Nemr, Pesticides pollution: classifications, human health impact, extraction and treatment techniques, Egypt. J. Aquat. Res. (2020) 207–220, https://doi.org/10.1016/j.ejar.2020.08.007. National Institute of Oceanography and Fisheries.
[24] R. Jayaraj, P. Megha, P. Sreedev, Review Article. Organochlorine pesticides, their toxic effects on living organisms and their fate in the environment, Interdiscipl. Toxicol. (2016) 90–100, https://doi.org/10.1515/intox-2016-0012. Slovak Toxicology Society.
[25] M.B. Bishop, C.B. Bishop, Photosynthesis and carbon dioxide fixation, J. Chem. Educ. (1987) 302–305, https://doi.org/10.1021/ed064p302.
[26] AR5 Climate Change 2013: The Physical Science Basis — IPCC (no date). Available at: https://www.ipcc.ch/report/ar5/wg1/(Accessed July 2, 2021).
[27] Z. Kriváсsy, M. Blazsó, D. Shooter, Primary organic pollutants in New Zealand urban aerosol in winter during high PM10 episodes, Environ. Pollut. 139 (2) (2006) 195–205, https://doi.org/10.1016/j.envpol.2005.05.018.
[28] D. Ji, et al., Characteristics of atmospheric organic and elemental carbon aerosols in urban Beijing, China, Atmos. Environ. 125 (2016) 293–306, https://doi.org/10.1016/j.atmosenv.2015.11.020.
[29] M. Blazsó, et al., Study of tropical organic aerosol by thermally assisted alkylation-gas chromatography mass spectrometry, J. Anal. Appl. Pyrolysis (2003) 351–369, https://doi.org/10.1016/S0165-2370(03)00082-2.
[30] R. Koppmann, Volatile organic compounds in the atmosphere. Volatile Organic Compounds in the Atmosphere, 2007, https://doi.org/10.1002/9780470988657. Blackwell Publishing Ltd.
[31] M.S. El-Shahawi, et al., An overview on the accumulation, distribution, transformations, toxicity and analytical methods for the monitoring of persistent organic pollutants, Talanta (2010) 1587–1597, https://doi.org/10.1016/j.talanta.2009.09.055. Elsevier B.V.
[32] Determination of Chemical Oxygen Demand of Wastewater : Pharmaceutical Guidelines (no date). Available at: https://www.pharmaguideline.com/2013/06/COD-test-waste-water-organic-pollution-determination.html (Accessed July 2, 2021).
[33] T. Damstra, Potential effects of certain persistent organic pollutants and endocrine disrupting chemicals on the health of children, J. Toxicol. Clin. Toxicol. (2002) 457–465, https://doi.org/10.1081/CLT-120006748.
[34] W.J. Criddle, G.P. Ellis, The separation of organic mixtures, in: Qualitative Organic Chemical Analysis, Springer US, 1967, pp. 17–20, https://doi.org/10.1007/978-1-4899-6383-3_3.
[35] S.J. Tinnemans, et al., Combining operando techniques in one spectroscopic-reaction cell: new opportunities for elucidating the active site and related reaction mechanism in catalysis, Catal. Today (2006) 3–15, https://doi.org/10.1016/j.cattod.2005.11.076.
[36] R. Kumar, V.B. Kumar, A. Gedanken, Sonochemical synthesis of carbon dots, mechanism, effect of parameters, and catalytic, energy, biomedical and tissue engineering applications, Ultrason. Sonochem. (2020) 105009, https://doi.org/10.1016/j.ultsonch.2020.105009. Elsevier B.V.
[37] L. Nizzetto, et al., Past, present, and future controls on levels of persistent organic pollutants in the global environment, Environ. Sci. Technol. (2010) 6526–6531, https://doi.org/10.1021/es100178f. American Chemical Society.
[38] A. Tiwari, et al., Fluorescent mantle carbon coated core-shell SPIONs for neuroengineering applications, ACS Appl. Bio Mater. 3 (7) (2020) 4665–4673, https://doi.org/10.1021/acsabm.0c00582.
[39] W.A. Battaglin, et al., Pharmaceuticals, hormones, pesticides, and other bioactive contaminants in water, sediment, and tissue from Rocky Mountain National Park, 2012–2013, Sci. Total Environ. 643 (2018) 651–673, https://doi.org/10.1016/j.scitotenv.2018.06.150.

[40] Q.Q. Li, et al., Persistent organic pollutants and adverse health effects in humans, J. Toxicol. Environ. Health (2006) 1987–2005, https://doi.org/10.1080/15287390600751447.
[41] D.C.G. Muir, P.H. Howard, Are there other persistent organic pollutants? A challenge for environmental chemists, Environ. Sci. Technol. (2006) 7157–7166, https://doi.org/10.1021/es061677a. American Chemical Society.
[42] S.N. Zulkifli, H.A. Rahim, W.J. Lau, Detection of contaminants in water supply: a review on state-of-the-art monitoring technologies and their applications, Sens. Actuators B (2018) 2657–2689, https://doi.org/10.1016/j.snb.2017.09.078. Elsevier B.V.
[43] N. Shahrubudin, T.C. Lee, R. Ramlan, An overview on 3D printing technology: technological, materials, and applications, Procedia Manuf. (2019) 1286–1296, https://doi.org/10.1016/j.promfg.2019.06.089. Elsevier B.V.
[44] H. Kodama, Automatic method for fabricating a three-dimensional plastic model with photo-hardening polymer, Rev. Sci. Instrum. 52 (11) (1981) 1770–1773, https://doi.org/10.1063/1.1136492.
[45] I. Karakurt, L. Lin, 3D printing technologies: techniques, materials, and post-processing, Curr. Opin. Chem. Eng. (2020) 134–143, https://doi.org/10.1016/j.coche.2020.04.001. Elsevier Ltd.
[46] T.S. Cheng, et al., 3D-printed metal electrodes for electrochemical detection of phenols, Appl. Mater. Today 9 (2017) 212–219, https://doi.org/10.1016/j.apmt.2017.07.005.
[47] D. Balram, K.Y. Lian, N. Sebastian, Ultrasound-assisted synthesis of 3D flower-like zinc oxide decorated fMWCNTs for sensitive detection of toxic environmental pollutant 4-nitrophenol, Ultrason. Sonochem. 60 (2020) 104798, https://doi.org/10.1016/j.ultsonch.2019.104798.
[48] V. Urbanová, M. Pumera, Uranium detection by 3D-printed titanium structures: towards decentralized nuclear forensic applications, Appl. Mater. Today 21 (2020) 100881, https://doi.org/10.1016/j.apmt.2020.100881.
[49] T. Vats, et al., Facile synthesis of pristine graphene-palladium nanocomposites with extraordinary catalytic activities using swollen liquid crystals, Sci. Rep. 6 (1) (2016) 1–11, https://doi.org/10.1038/srep33053.
[50] M. Zafir Mohamad Nasir, et al., 3D-printed electrodes for the detection of mycotoxins in food, Electrochem. Commun. 115 (2020) 106735, https://doi.org/10.1016/j.elecom.2020.106735.
[51] K.P.A. Kumar, et al., Metal-plated 3D-printed electrode for electrochemical detection of carbohydrates, Electrochem. Commun. 120 (2020) 106827, https://doi.org/10.1016/j.elecom.2020.106827.
[52] J. Karpińska, U. Kotowska, Removal of organic pollution in the water environment, Water (Switzerland) (2019) 2017, https://doi.org/10.3390/w11102017. MDPI AG.
[53] J. Rivera-Utrilla, et al., Activated carbon modifications to enhance its water treatment applications. An overview, J. Hazard Mater. (2011) 1–23, https://doi.org/10.1016/j.jhazmat.2011.01.033. Elsevier.
[54] P.V. Nidheesh, et al., A review of integrated advanced oxidation processes and biological processes for organic pollutant removal, Chem. Eng. Commun. (2021), https://doi.org/10.1080/00986445.2020.1864626.
[55] M. Chawla, R. Kumar, P.F. Siril, High catalytic activities of palladium nanowires synthesized using liquid crystal templating approach, J. Mol. Catal. Chem. 423 (2016) 126–134, https://doi.org/10.1016/j.molcata.2016.06.014.
[56] S. Dutt, R. Kumar, P.F. Siril, Green synthesis of a palladium-polyaniline nanocomposite for green Suzuki-Miyaura coupling reactions, RSC Adv. 5 (43) (2015) 33786–33791, https://doi.org/10.1039/c5ra05007c.

[57] X. Xia, et al., Microalgal-immobilized biocomposite scaffold fabricated by fused deposition modeling 3D printing technology for dyes removal, ES Mater. Manuf. (2020), https://doi.org/10.30919/esmm5f706.
[58] H. Zhao, X. Song, H. Zeng, 3D white graphene foam scavengers: vesicant-assisted foaming boosts the gram-level yield and forms hierarchical pores for superstrong pollutant removal applications, NPG Asia Mater. 7 (3) (2015), https://doi.org/10.1038/am.2015.8. pp. e168–e168.
[59] Z. Shi, et al., Renewable metal-organic-frameworks-coated 3D printing film for removal of malachite green, RSC Adv. 7 (79) (2017) 49947–49952, https://doi.org/10.1039/c7ra10912a.

CHAPTER

10

Resource management using 3D printing technology

Pritam Dey, and Srimanta Ray

Department of Chemical Engineering, National Institute of Technology Agartala, Agartala, Tripura, India

Abbreviations

AM Additive manufacturing
ASTM American society for testing and manufacturing
CAD Computer aided design
LDPE Low density polyethylene
SLS Selective laser sintering
SLM Selective laser melting
STL Standard tessellation language

10.1 Introduction

Historically, business models have revolved around the manufacturing of products and their utility by the consumers but not much attention is paid to the utilization of the wastes produced. A systemic approach to economic development designed to benefit businesses, society, and the environment by keeping the materials used in a closed-loop is known as *Circular Economy* [1]. In a circular economy, a product after being created and used, rather than throwing away the waste, its materials are recovered to be used yet again in creating new products. Three-dimensional Printing or Additive Manufacturing (AM) has huge potential to cut back the intensive consumption of both energy and materials. ASTM International defines AM as the "process of joining materials to make objects from 3D model data, usually layer upon layer, as opposed to subtractive manufacturing methodologies" [2]. Some of the most important benefits of 3D printing are its efficient use of materials and minimal waste production when compared with traditional, subtractive manufacturing methods.

https://doi.org/10.1016/B978-0-323-99861-1.00009-6

10.1.1 History of 3D printing

In 1981, Hideo Kodama completed his work on the development of a laser cured rapid prototyping technique. Over the subsequent three decades, this invention was further exploited due to the introduction of stereolithography in 1984 [3]. The very first 3D printer was invented by Chuck Hull of 3D Systems in 1987. Following this, some noteworthy innovations like selective laser sintering (SLS) and selective laser melting (SMS) techniques came into the picture. Between 1990 and 2000, some efficient but expensive 3D printing techniques were developed. However, when the patent on 3D printing expired in 2009, the prices got reduced drastically, creating more buzz in the technological world [4].

10.1.2 Salient features of 3D printing

Three-dimensional printing or AM is the method of manufacturing a 3D object from scratch using a computer-aided design (CAD) model. A standard tessellation language (STL) file is usually required by many 3D printers to obtain the data required for printing. An STL file is a simple and portable file format required for defining the solid geometry of the 3D printable object. An STL file models the overall surface of the object in the form of asymmetric geometrical shape (usually triangle) in which each edge and vertex of the shape share the edges and vertices with their neighboring shape to build the complete 3D structure. The quality and precision with which the 3D object gets printed rely on the resolution of the STL file [4]. A variety of processes are utilized in which material is deposited, joined, or solidified under a computer-controlled system to construct a solid 3D object. In 3D printing, the precursor materials are joined together in a layer-by-layer fashion. A variety of materials can be used as construction material for 3D printing. The three broad categories of 3D printing technology are sintering, melting, and stereolithography. Sintering is a technology where the material is heated to high temperatures, but not to the point of melting [3]. This is done to generate high-resolution items. Melting methods under 3D printing comprise electron beam melting, powder bed fusion, and direct energy deposition [5]. In these methods, the construction material is melted and given the desired shape using electric arcs, lasers, or electron beams as the energy source. Stereolithography utilizes photopolymerization to create the desired components [5]. In this method, the light of fixed wavelength interacts with the construction material in a well-defined condition for curing and solidifying a layer-by-layer cross-section of the desired product.

Thus the objective of this chapter is to provide a brief but insightful review of 3D printing techniques used in resource management, with a focus on the innovations in water treatment. Discussion on the important methods and principles employed in 3D printing, materials utilized, and its current state and applications in various water treatment industries is also presented. The chapter also highlights the present research gaps and challenges encountered in adopting this technology.

10.2 The significance of 3D printing

10.2.1 3D printing (or AM) versus traditional manufacturing methods

The 3D printing technology has drastically reduced the extra expenditures that are incurred within the course of manufacturing an item from its prototype [6]. The production of customized and tailored products has been an uphill task for the industries due to the high costs of materials and manufacturing processes. On the contrary, 3D printing can create small quantities of customized products at relatively low costs. Owing to the benefits of high precision, flexibility in design, minimum material wastage, and the option of constructing highly customized products, there has been a rise in the acceptance of 3D printing technology among manufacturers. However, the issues of weakened mechanical properties and rise in the anisotropic behaviors of the 3D printed objects have limited the use of this technology in large-scale manufacturing processes. Thus, it has become imperative to improve 3D printing patterns for minimizing the mechanical flaws and anisotropy in the final product.

The benefits linked with 3D printing technology are:

- *Efficient and effective construction of complex geometries.* 3D printing technology permits the straightforward creation of bespoke geometric parts where added complexity comes at no extra cost. In some instances, 3D printing is cheaper than subtractive production methods as no extra material is employed.
- *Affordable manufacturing costs.* As there is no requirement for molds for manufacturing products, the overall cost of the manufacturing process is relatively low.
- *Ease of customization.* As 3D printing is predicated upon CAD-based designs, any change in the product geometry or shape is possible without impacting the production cost.
- *Most suitable for swift prototyping.* As the technology permits product construction in small lots and modular fabrication, 3D printing is good for swiftly creating a prototype, which implies that the prototypes will be produced in no time.
- *Products with desired characteristics.* Because a layer-by-layer construction of the product takes place in 3D printing, it is possible to construct a product using an engineering material with desired characteristics.

Whereas, the demerit associated with 3D printing are:

- *Lower mechanical strength of the products.* While parts made of metals have excellent mechanical properties, many 3D printed parts are more brittle than those created by traditional manufacturing techniques. This can be because the parts are built up layer-by-layer, which reduces the strength by 10%–50%.
- *Amplified cost for a high volume of products.* Literature suggests that when making an on-the-spot comparison for similar or analogous parts, 3D printing is costlier compared to traditional machining or injection molding while manufacturing products in high volumes.
- *Rare inadequacies inaccuracy.* Some desktop printers with low tolerances may generate parts that might vary from the provided designs. While post-processing may fix the glitch, this shows that 3D printed products might not always be akin to each other.
- *Requirement of postprocessing steps.* Most 3D printed components necessitate a postprocessing step, like sanding or smoothing, removal of support structures, heat treatment to attain specific material properties, and or final machining.

10.2.2 Scopes and challenges of 3D printing

Three-dimensional printing or AM could boost global "resource yield" by generating more products out of a fixed quantity of a given resource and significantly reducing the generated waste. This may prove pivotal in narrowing the gap between the availability and usage of nonrenewable resources [7].

Even so, before extending 3D printing or AM to a wider variety of manufacturing processes, a few remaining issues must be tackled. Certain issues are more prominent in some 3D printing methods or materials. Three-dimensional printing methods such as powder-bed (SLS or SLM) have a high resolution, resulting in high energy consumption. Three-dimensional printing of a component, for example, takes longer than some conventional methods such as welding or injection molding. In comparison to inkjet printing and fused deposition modeling, the powder bed process and stereolithography take longer. The long processing time and higher cost of 3D printing are the key barriers to mass production of any repetitive parts that can be easily produced using other traditional methods in a fraction of the time and cost.

Despite the time and expense of 3D printing output, there are other major challenges associated with the design which need to be evaluated for future progress.

Some of those challenges are:

1. Layer-by-layer appearance
2. From conception to completion, there are a wide range of styles.
3. Maintaining homogeneity of the feedstock throughout the process
4. Mechanical properties and tensile strength
5. Porosity
6. Void formation

10.2.3 Diverse applicability of 3D printing technology

The flexible nature of 3D printing technology has resulted in its utilization across a variety of manufacturing divisions. In place of constructing a product from a variety of components, 3D printing permits the manufacturing of a single product, decreasing the production time and wastage of raw materials. The automotive sector has accepted the 3D printing technology because of its ability to lower the weight of the component. Three-dimensional printing technology is in demand across the aerospace industry to manufacture lightweight and complex engineered aircraft parts. Also, with 3D printing quick prototyping of customized parts for tests or quality assessment can be achieved [8]. With the help of this, the problem of the unavailability of certain specific parts can be overcome. The rail transportation industry has employed 3D printing for creating customized parts, such as armrests for drivers and housing covers for train couplings, etc. [9]. The medical sector has embraced 3D printing technology for the production of medical-grade implants and small devices [9]. For instance, miniaturized hearing devices can be created in no time with the help of a digital file that matches with a scan of the internal ear structure of a patient. The swift manufacturing, ease of designing, and ability to create custom-made components have catapulted 3D printing technology in the robotics industry. Not just in mainstream industries, 3D printing is additionally revolutionizing the industries involved in sustainable development and environmental protection.

10.3 Current applications of 3D printing in water treatment

10.3.1 3D printed membranes

The advancements in the field of 3D printing also pushed other areas of science and technology to develop in tandem. Currently, basic as well as applied sciences have already enjoyed the benefits of 3D printing technology and its ever-increasing growth. Only in recent years, the fields of environmental science and engineering have embraced this technology [5]. The fabrication of various membrane modules for water treatment purposes is such an area that has been benefited the most from the advancements in 3D printing technology. The application of 3D printing technology has started with membrane-based studies, however, the fabrication of 3D printed membranes has evolved as a new field in environmental research. Membrane science and technology have become imperative in water treatment owing to the enhanced removal ability and improved energy efficiency. The necessity for highly efficient membranes also demanded swift and accurate fabrication and hence 3D printing technology has come into the picture. Precise and accurate fabrication of membranes is additionally essential for the water-energy nexus. The 3D printing technology has made it possible to fabricate robust and ultrathin membranes which retain identical and well-organized pores for efficient water treatment. This is made possible by the layer-by-layer printing of the membrane which is an advantage of the 3D printing technology [7]. As a result, the ion-selective nature of the membrane can be enhanced and can be easily controlled during the fabrication process.

The conventional membrane fabrication processes include nonsolvent-induced phase separation, thermally induced phase separation, and vapor-induced separation [10]. One of the major disadvantages linked with these processes is the requirement of tremendous amounts of solvents which in turn produce harmful byproducts and wastewater, resulting in huge environmental impacts. The monomers for water treatment membranes like N-methyl-2-pyrrolidone, N, N-dimethylformamide, and N, N-dimethylacetamide are highly toxic and have severe health hazards [5]. Furthermore, it has been estimated that over 50 billion liters of wastewater are generated per year during the conventional membrane fabrication process [11]. Consequently, the generation of minimum waste and low chemical requirement, which improve the economics of the process, make 3D printing of water treatment membrane an important technology that needs to be fully utilized.

Three-dimensional printing is currently applied for the fabrication of membrane modules, surfaces, parts, and support layers [10]. There is still a long way to go in terms of establishing membrane fabrication technology for the large-scale production of water treatment membranes. The reason is that 3D printing technology is itself evolving and developing. Nevertheless, 3D printing has helped in transforming the designing of membrane parts and modules, lowering the material and energy consumption and wastage, and reducing the dependence on harmful chemicals during the fabrication process. Hence, to speed up the developments in the fabrication of 3D printed membranes key aspects of technology and manufacturing processes should be identified and further research in this area is required.

10.3.2 Methods used for fabrication of 3D printed membranes

Methodologies involved in the fabrication of membranes via 3D printing may be classified depending on the nature of the material utilized and the method of manufacturing. Close to 50 such 3D printing technologies have been identified so far. All these techniques have been further categorized by the ASTM 52900:2015 standard into seven classes (Fig. 10.1) [12]. The specific 3D printing techniques that are identified for the fabrication of membranes include powder bed fusion technologies, inkjet-based fabrication methods, binder jetting, vat photopolymerization, material extrusion, sheet lamination, and direct energy deposition. An analogous four-step methodology that begins with building a CAD-based model is employed in all these methods. In the second step, the model is converted into a 3D printer readable interface, using an STL file system. Following that the model is sliced into multiple 2D layers which are fabricated in a layer-by-layer manner using a 3D printer to construct the

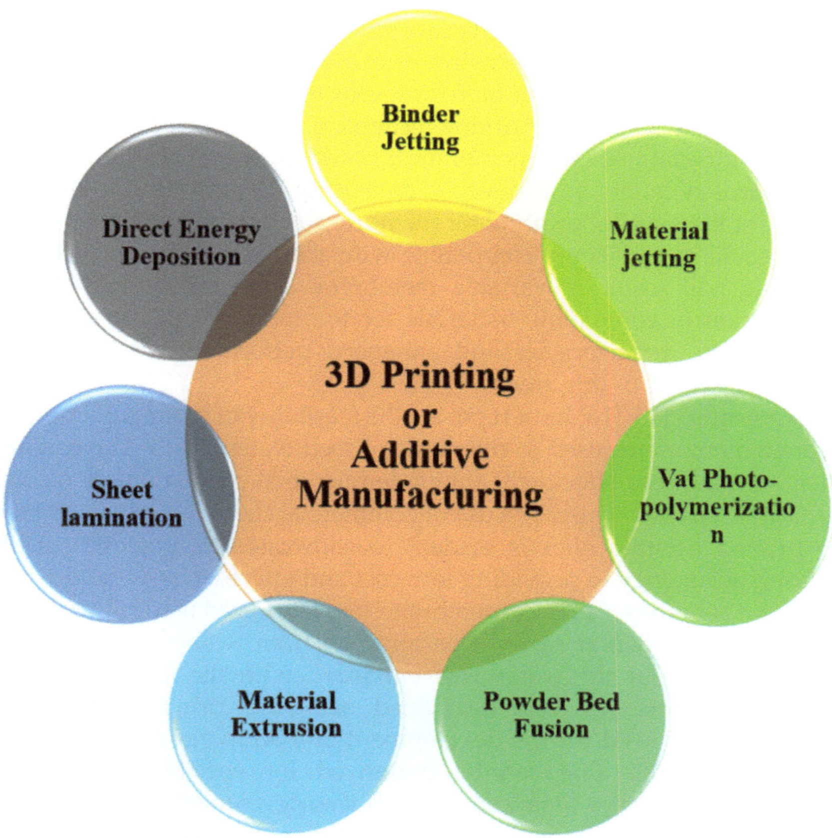

FIGURE 10.1 Classification of various 3D printing processes.

final product. More than often post-processing of the final product is done to achieve the desired finish and quality. Materials generally utilized for 3D printing include various thermoplastic materials like polypropylene and polystyrene. Along with these Nylon 618, acrylonitrile, butadiene, styrene, Nylon 680, polylactic acid, and Nafion 117, various photopolymeric resins, ceramics, metals, and their alloys are also employed [13].

Table 10.1 presents the advantages and disadvantages of various 3D printing processes involved in the manufacturing of membranes for water purification.

10.3.3 Advances of 3D printing in water and waste water treatment

Advancements in the field of water and wastewater treatment have led to the development of a variety of techniques in which fabrication of 3D printed membranes is a recent technique [14]. In this regard, conventional water treatment methods are being replaced with membrane technologies. Since 3D printing supports layer-by-layer fabrication and designs with multiple levels of hierarchy, the membranes fabricated by 3D printing technology shall be more cost-effective than traditionally fabricated membranes. The 3D printing technology has shown prominence in fabricating ceramic-based membranes; however, submicron level membrane porosity cannot be achieved to date using 3D printing [15]. Another constraint in using 3D printing for membrane development is that only a few specific materials are compatible with the technology and all such materials will result in variable membrane characteristics. Thus, the introduction of compatible materials for the fabrication of membranes with desired characteristics and enhanced performance is the future scope of research in this area.

Akowanou et al. [16] have reported the feasibility of various 3D printed ceramic membranes used in membrane-based water filters. The ceramic filter was found to effectively eliminate microbial population as well as turbidity from raw water. The use of commercial clay for constructing the 3D printed ceramic filter is evident to showcase the potential of 3D printing technology in providing low-cost and effective household water treatment solutions. Ceramic membranes are fabricated using 3D printing technology by Chen et al. for separating oil from water [17]. The separation of oil from the water was enhanced with the combination of aluminum borate whiskers developed using 3D printing. Thus, 3D printing has allowed for the development of a wide variety of complex internal structures that can be optimized for specific applications. Another oil and water separation system was developed by Li et al. [18]. In this system, a super-hydrophilic membrane, printed using a direct inkjet 3D printer is present which can efficiently separate oil from water. Additionally, a 3D printed ceramic mesh also performed well in the

TABLE 10.1 Advantages and disadvantages of various 3D printing processes for membrane manufacturing.

Process	Advantages	Disadvantages
Binder jetting	Numerous potential materials are available. Making metal casting molds out of ceramic is easier. Layer fabrication includes support structures by default.	A texture that is rough or grainy lacks strength. To remove moisture or increase strength, post-processing is required.
Direct energy deposition	High rate of material usage. Repair and add-on features have high efficiency. Wear-resistant metals are deposited in thin layers on components.	Part complexity ranges from low to medium. Surface quality and resolution are poor. Dimensional precision is poor. Materials for manufacture are limited.
Material extrusion	The entry-level machines are inexpensive. A wide range of raw materials is accessible, flexible, and simple to personalize	Low precision and a long construction time. Lacking the ability to create sharp exterior corners. The printed part's anisotropic nature.
Material jetting	Minimum wastage of material. High precision and resolution. Various materials and colors are used.	Thin and tiny features may be harmed by post-processing. Support materials can't be recycled, thus they're thrown away.
Powder bed fusion	Polymer powder does not require any additional support. Metal powder and polymer powder can both be recycled. Part complexity is high, and the materials used are diverse. Metals require high precision and resolution.	Polymer ends with a rough surface finish. The rate of construction is relatively slow. Expensive machines are only used for small to medium pieces.

Continued

TABLE 10.1 Advantages and disadvantages of various 3D printing processes for membrane manufacturing.—cont'd

Process	Advantages	Disadvantages
Sheet lamination	High-speed fabrication. There are no support structures required. Internal stress and low warping. It is possible to use multiple materials and colors.	Support stuck in internal cavities is difficult to remove. Thermal cutting emits toxic gases. As a result of the laser's heat, there's a chance that the lamination will deform.
Vat photo-polymerization	High precision and resolution. Good surface finish. High-speed fabrication. Low-imaging specific energy. It is possible to use multiple materials and colors.	Need support to remove moisture or increase strength, post-processing is required. Involve post-curing for enhanced strength.

system, after replacing the hydrophilic membrane. Such studies have shown the potential of 3D printed separation membranes that can become well-established in the coming future.

Beyond membrane fabrication for water treatment, the 3D printing technology has shown its potential development of photocatalysts and micro-devices for water remediation. Martin de Vidales et al. [19] have synthesized a titanium dioxide-based photocatalyst impregnated in a 3D printed low-density polyethylene (LDPE) mesoporous mesh. This photocatalyst was used for the removal of ofloxacin, an emerging organic pollutant, from synthetic wastewater. Titanium dioxide catalyst relies on ultraviolet radiation for its activation. The 3D printed LDPE mesh floats on water, while the photocatalyst housed in it gets activated in the presence of sunlight and carries out the photocatalytic degradation of ofloxacin. The photocatalytic efficiency was found to have improved with an increase in the availability of active sites, which was made possible by the 3D printing technology. Bernasconi et al. [20] have developed a microrobotic device using 3D printing, which was reported to have both photocatalytic as well as antimicrobial properties. Their microrobots were constructed using a 3D stereolithography method followed by a wet metallization technique. This combined method of fabricating micro-device was reported to be feasible for going forward to the production line because many devices can be metalized at a time. Both the floating photocatalyst and microrobotic device are novel innovations for providing effective low-cost water treatment solutions, made possible due to the advancements in the field of 3D printing technology.

10.4 Current trending fields where 3D printing is employed for water treatment and water quality analysis

10.4.1 In aquaculture

In the year 2016, over 80 million tons of fish were produced via fish farming or aquaculture [21]. In recent times the worldwide expansion of fish farms is evident to fulfill the growing demand for fish. As a result, the science of raising fish or pisciculture has also seen many technological advancements. It is important to control the water quality, pH, suspended and dissolved solids, etc., of commercial fisheries similar to those of open water systems. The evolution in 3D printing technology and recent upgrades in aquaculture have resulted in the development of innovative and easy-to-use water quality testing and monitoring systems [22]. Such systems can send notifications remotely for the farm which has made monitoring water quality consistent and hassle-free. Such innovations are being considered to make aquaculture or fish farming more cost-effective in the coming years.

A robotic fish was recently printed by researchers from the Massachusetts Institute of Technology that can precisely mimic the motions and movements of an actual fish [23]. Technology like this might provide prospects for further research in order to completely understand the behaviors of aquatic species in both their natural and artificial environments. Such research shall aid in making improvements in the captive conditions of the fish species which shall further prove beneficial for fish production.

A very different aspect of 3D printing consists of the assembly of structured seaweed. Seaweeds are raised in water bodies in specific arrangements for proper aeration, nutrients, water, and sunlight availability. In this regard, algal production holds the promising potential to provide food as well as fuel in the coming future. Algae have been cultured to manufacture environmentally friendly and inexpensive materials that can be used as starting material for constructed 3D printed medical implant devices. The process of producing human tissue and organs is additionally already on the agenda, algae may one day lead to the generation of life-saving procedures.

10.4.2 Desalination to create potable water

A small fraction of the global world is fit for use and is used in agriculture, industrial, and household purposes. It has been predicted that the global population will escalate by 40%–50% in the next half a century, which shall lead to a decline in water availability [24]. Membrane-based desalination of surface water is considered the most effective water treatment technology. However, this method is very costly because of the inherent cost of the membrane employed in the process. By creating water treatment membranes using 3D printing technology, soon there shall be an effective expansion in the global availability of potable water [7]. Recent studies have concluded that 3D printed ceramic membranes are able to effectively mitigate the issue of membrane fouling that usually occurs during water treatment [25]. As mentioned earlier, 3D printed membranes are expected to be more efficient and cost-effective than conventionally manufactured membranes. Also, such membranes are linked with far lower environmental impacts.

10.4.3 Efficient water analysis devices

Contaminated water is still a very prominent cause of death in developing and underdeveloped countries. Poor water quality sometimes results in life-threatening illnesses. To remedy this problem, an efficient assessment of the quality of potable water is highly pertinent. For these reasons, the rural communities are required to be equipped with technologies to check water quality remotely and in a cost-effective manner. Researchers have developed a facile 3D printed device for remotely assessing the surface water quality [26]. The device employs several low-resolution microcameras to record color changes in water for performing a colorimetric assay to determine the water quality. The device can be connected wirelessly for data gathering, processing, and storage. Moreover, the determination of heavy metals like cadmium and lead in wastewater has been achieved using fluorescent-based assay on a 3D printed lab-on-chip device [27]. Calderilla et al. [28] have fabricated a 3D printed disc for solid-phase extraction of chromium(VI) from industrial wastewater and its subsequent quantitative measurement in-vitro. Further advancements in the manufacture of 3D printed devices and analyzers shell revolutionized the way we analyze water or wastewater.

10.4.4 Water filtration and purification systems

The analysis of water for determining various contaminations is incredible, but the advancement in technology that permits us to get rid of the detected contaminations is groundbreaking. While the fabrication of efficient water purification or filtration systems is possible now, the capability to do so in a cost-effective and swift manner implies that such systems shall reach the inaccessible areas and neediest of communities. The 3D printing-based electrospinning process can develop nanofiber membranes for advanced water filters [9]. Products such as the "Naked Filter" use this innovative technology to create filters 70%–80% less porous than traditional filters, making them far more effective in filtering out pathogenic microbes such as *E. coli* or *Salmonella* [29].

10.4.5 Efficient wastewater treatment

According to a United Nations report, over four-fifth of global water is not collected or treated [30]. Countries like China and India are developing very fast and hence there's a necessity for the water infrastructure of these countries to develop rapidly to stay on pace with the growing population. Wastewater treatment is often a costly affair in such countries where the 3D printing technology comes in handy in reducing the process cost. Three-dimensional printing is being utilized in water applications like desalination, water and wastewater treatment [13,14]. Around 20% of the world's current population faces the issue of water scarcity and the number is expected to cross 1.8 billion by the next 5 years or so [24]. As a result of the growing population and increasing water scarcity, our water systems are becoming more and more strained. The 3D printing technology is facilitating to make groundbreaking innovations to boost our surroundings and transform communities in need. From generating potable water to treating wastewater to aquaculture, 3D printing technology is revolutionizing the way we obtain water for survival. Groundbreaking 3D printing technology shall open the chance to remodel water infrastructures around the globe.

10.4.6 Metal recovery from wastewater

The occurrence of metals (and heavy metals) in wastewater streams has been a worldwide persistent problem. Such metals are biomagnified in ground and surface water affecting both human and animal health [31]. Moreover, many valuable metals like gold, copper, silver, aluminum, etc., are usually lost in the wastewater because of the unavailability of suitable recovery methods. Electrochemical recovery has evolved as a suitable metal recovery method in recent times [31]. Three-dimensional printing technology has the potential to facilitate the electrochemical recovery of metals by fabricating efficient membranes, electrodes, and other parts comprising complex geometries. Recently, Lahtinen et al. [32] have developed a 3D-printed polymeric scavenger containing meshed structure for selective gold recovery from electronic liquid waste. The gold ions can be effectively and quantitatively removed from the scavenger structure using a simple elution method. Similarly, 3D-printed objects can be effectively utilized for the recovery of other metals due to high selectivity and enhanced cross-sectional area.

10.4.7 Recovery and management of organics from waste streams

Many valuable organic compounds are lost through industrial wastewater. Recovery of such organic compounds will lead to economic gains and will lessen the burden on natural resources. Moving bed bioreactors are popular for treating wastewater, owing to their simple design and superior organic removal efficiency [33]. Three-dimensional printing technology is looked onto for manufacturing such bioreactors in no time with enhanced surface properties and complex geometries. The advancements in the field of membrane fabrication and water remediation using 3D printing technology may prove beneficial in recovering selective organic compounds from wastewater. Also, 3D printing technology has shown ways to manufacture objects for resource recovery in a sustainable fashion. In this regard, Maciel et al. [34] have developed polymerisable ionic liquids immobilized in 3D printed solid shapes which can be used for the recovery of organic solvents from industrial wastewater. The 3D printing technology has led to an effective reduction in solvent consumption and improvements in the synthesis process. Moreover, organic resources recovered from waste can be effectively recycled using 3D printing. An example is the development of lignin-based biocomposite using 3D printing technology [35]. Lignin recovered from waste forest biomass was used as a filer in the polylactic acid matrix,and different shapes were developed using facile and accurate 3D printing techniques. Thus, 3D printing technology can revolutionize material recovery from waste and may become an essential waste management tool.

10.5 Summary

The technological advancements in the field of 3D printing have revolutionized the production and manufacturing sector due to its accuracy, customizability, minimal waste generation, and sustainable fabrication outcomes. Polymers are the most common materials developed for rapid prototyping and swift manufacturing of 3D printed components. As a result, water treatment, like other fields of research, has effectively accepted and implemented this technology for the production of better membranes, eliminating high carbon emissions generated by fabrication, and preventing significant quantities of solvent wastage. The development of novel membrane modules and the fabrication of efficient and cost-effective membranes are the subjects of important research involving 3D printing technology. Printing nanomaterials on the surface of membranes using advanced 3D printing techniques has resulted in high selectivity, increased porosity, and improved fouling resistance. The 3D printed polymeric membrane supports have recently been implemented for water treatment and management. Furthermore, with the growing water scarcity and rise in global population the pressure on our existing water infrastructure is increasing. Technological advancements such as 3D printing in aquaculture, water desalination, water purification;

development of low-cost water testing kits, and material and resource recovery tools are assisting in the development of ground-breaking technology to enhance our climate and transform underserved communities. Three-dimensional printing technology has also proven efficient in recovering valuable metals and organic compounds from wastewater. However, more research and innovations are necessary to make 3D printing technology more affordable and easily available in every corner of the world.

References

[1] M. Geissdoerfer, P. Savaget, N.M. Bocken, E.J. Hultink, The Circular Economy—A new sustainability paradigm? J. Clean. Prod. 143 (2017) 757—768.

[2] ASTM F2792-10, Standard Terminology for Additive Manufacturing Technologies, 2010.

[3] K.R. Piyush, R. Kumar, 3D printing of food materials: a state of the art review and future applications, Mater. Today Proc. 33 (2020) 1463—1467.

[4] V. Gupta, P. Nesterenko, B. Paull, 3D Printing in Chemical Sciences: Applications across Chemistry, RSC Publishers, London, 2019.

[5] N. Shahrubudin, T.C. Lee, R. Ramlan, An overview on 3D printing technology: technological, materials, and applications, Procedia Manuf. 35 (2019) 1286—1296.

[6] J.-Y. Lee, J. An, C.K. Chua, Fundamentals and applications of 3D printing for novel materials, Appl. Mater. Today 7 (2017) 120—133.

[7] T. Campbell, C. Williams, O. Ivanova, B. Garrett, Could 3D printing change the world? technologies, potential, and implications of additive manufacturing, Atlantic Council (2011) 1—17.

[8] P.S. Ujeniya, N.V. Rachchh, A review on Manufacturing, Machining, and Recycling of 3D printed composite materials, IOP Conf. Ser. Mater. Sci. Eng. 653 (2019) 012024.

[9] T.D. Ngo, A. Kashani, G. Imbalzano, K.T.Q. Nguyen, D. Hui, Additive manufacturing (3D printing): a review of materials, methods, applications, and challenges, Compos. B Eng. 143 (2018) 172—196.

[10] N. Yanar, P. Kallem, M. Son, H. Park, S. Kang, H. Choi, A New era of water treatment technologies: 3D printing for membranes, J. Ind. Eng. Chem. (Seoul, Repub. Korea) 91 (2020) 1—14.

[11] H.H. Wang, J.T. Jung, J.F. Kim, S. Kim, E. Drioli, Y.M. Lee, A novel green solvent alternative for polymeric membrane preparation via non-solvent-induced phase separation (NIPS), J. Membr. Sci. 574 (2019) 44—54.

[12] ASTM 52900-15, Additive Manufacturing — General Principles — Terminology, 2015.

[13] H.A. Balogun, R. Sulaiman, S.S. Marzouk, A. Giwa, S.W. Hasan, 3D printing and surface imprinting technologies for water treatment: a review, J. Water Proc. Eng. 31 (2019) 100786.

[14] M.N. Nadagouda, M. Ginn, V. Rastogi, A review of 3D printing techniques for environmental applications, Curr. Opin. Chem. Eng. 28 (2020) 173—178.

[15] M.N. Issac, B. Kandasubramanian, Review of manufacturing three-dimensional-printed membranes for water treatment, Environ. Sci. Pollut. Res. 27 (2020) 36091—36108.

[16] A.V.O. Akowanou, H.E.J. Deguenon, L. Groendijk, M.P. Aina, B.K. Yao, P. Drogui, 3D-printed clay-based ceramic water filters for point-of-use water treatment applications, Prog. Addit. Manuf. 4 (2019) 315—321.

[17] Z. Chen, D.W. Zhang, E. Peng, J. Ding, 3D-printed ceramic structures with in situ grown whiskers for effective oil/water separation, Chem. Eng. J. 373 (2019) 1223–1232.
[18] X. Li, H. Shan, W. Zhang, B. Li, 3D printed robust superhydrophilic and underwater superoleophobic composite membrane for high efficient oil/water separation, Sep. Purif. Technol. 237 (2020) 116324.
[19] M.J. De Vidales, A. Nieto-Márquez, D. Morcuende, E. Atanes, F. Blaya, E. Soriano, F. Fernández-Martínez, 3D printed floating photocatalysts for wastewater treatment, Catal. Today 328 (2019) 157–163.
[20] R. Bernasconi, E. Carrara, M. Hoop, F. Mushtaq, X.Z. Chen, B.J. Nelson, S. Pane, C. Credi, M. Levi, L. Magagnin, Magnetically navigable 3D printed multifunctional micro-devices for environmental applications, Addit. Manuf. 28 (2019) 127–135.
[21] U.R. Sumaila, C. Bellmann, A. Tipping, Fishing for the Future: an overview of challenges and opportunities, Mar. Pol. 69 (2016) 173–180.
[22] K. Yue, Y. Shen, An Overview of Disruptive Technologies for Aquaculture. Aquaculture Fisheries, 2021.
[23] Y. Horii, K. Inoue, S. Nishikawa, K. Nakajima, R. Niiyama, Y. Kuniyoshi, Physical reservoir computing in a soft swimming robot, in: ALIFE 2021 – the 2021 Conference on Artificial Life, MIT Press, 2021.
[24] J.I. Uitto, Population, food, and water in the 21st century, in: Population Problems, Routledge, 2020, pp. 93–110.
[25] Z. Lyu, T.C.A. Ng, T. Tran-Duc, G.J.H. Lim, Q. Gu, L. Zhang, et al., 3D-printed surface-patterned ceramic membrane with enhanced performance in crossflow filtration, J. Membr. Sci. 606 (2020) 118138.
[26] G.S. Luka, E. Nowak, J. Kawchuk, M. Hoorfar, H. Najjaran, Portable device for the detection of colorimetric assays, R. Soc. Open Sci. 4 (2017) 171025.
[27] E. Mattio, F. Robert-Peillard, L. Vassalo, C. Branger, A. Margaillan, C. Brach-Papa, et al., 3D-printed lab-on-valve for fluorescent determination of cadmium and lead in water, Talanta 183 (2018) 201–208.
[28] C. Calderilla, F. Maya, V. Cerdà, L.O. Leal, 3D printed device for the automated preconcentration and determination of chromium (VI), Talanta 184 (2018) 15–22.
[29] L.D. Tijing, J.R.C. Dizon, I. Ibrahim, A.R.N. Nisay, H.K. Shon, R.C. Advincula, 3D printing for membrane separation, desalination and water treatment, Appl. Mater. Today 18 (2020) 22.
[30] D.N. Chakkaravarthy, T. Balakrishnan, Water scarcity-Challenging the future, Int. J. Agric. Environ. Biotechnol. 12 (2019) 187–193.
[31] W. Jin, Y. Zhang, Sustainable electrochemical extraction of metal resources from waste streams: from removal to recovery, ACS Sustain. Chem. Eng. 8 (12) (2020) 4693–4707.
[32] E. Lahtinen, L. Kivijärvi, R. Tatikonda, A. Väisänen, K. Rissanen, M. Haukka, Selective recovery of gold from electronic waste using 3D-printed scavenger, ACS Omega 2 (10) (2017) 7299–7304.
[33] O. Elliot, S. Gray, M. McClay, B. Nassief, A. Nunnelley, et al., Design and manufacturing of high surface area 3D-printed media for moving bed bioreactors for wastewater treatment, J. Contemp. Water Res. Edu. 160 (2017) 144–156.
[34] V.G. Maciel, D.J. Wales, M. Seferin, V. Sans, Environmental performance of 3D-Printing polymerisable ionic liquids, J. Clean. Prod. 214 (2019) 29–40.
[35] M. Tanase-Opedal, E. Espinosa, A. Rodríguez, G. Chinga-Carrasco, Lignin: a biopolymer from forestry biomass for biocomposites and 3D printing, Materials 12 (18) (2019) 3006.

CHAPTER

11

Sustainability of 3D printing in industry 4.0: A brief review

Aditya Sachdeva, Rashi Agrawal, Chetan Chaudhary, Darshit Siddhpuria, Deepanshu Kashyap, and Seim Timung

School of Engineering, University of Petroleum and Energy Studies, Dehradun, Uttarakhand, India

11.1 Industry 4.0—past, present, and future

The first Industrial Revolution gave rise to the production of goods through heavy machinery that was powered by steam. This remarkable change took place between 1760 and 1840 that had a major consequence on the produced textile goods, metallurgy, and agriculture industries. This change affected the nearby population by creating larger job opportunities. The second Industrial Revolution took place in the period between 1871 and 1914. This period saw the developments in infrastructures and improvements in railroad, electricity lines, and telegraph networking, for quicker transfer of people, and faster exchange of ideas. These factors led to higher production rates and the availability of goods in the market. However, the period also saw an increase in unemployment as many of the laborers working in factories were quickly replaced by continuously operating machines [1]. The third Industrial Revolution that happened in the late 20th century introduced the integration of computers and miniaturized electronic chips in the industrial production lines. This period was marked as the era of the digital revolution due to its major shift in the technology toward Information and Communication Technologies (ICT) and allowed the shifting of the mechanical and analog mode of electronics to that of the digitized mode of electronics. This period saw the mass production along with the extensive use of digital logic, integrated circuits (IC), MOSFETs (MOS transistors), chips, microprocessors, wireless communication phones, and the internet.

3D Printing Technology for Water Treatment Applications
https://doi.org/10.1016/B978-0-323-99861-1.00010-2

These technological innovations have altered the traditional way of production, business tactics, and maintaining records [2]. Moreover, Industry 4.0 refers to a shift in automated machines and the exchange of information related to production technology and operations through cyber-physical systems (CPS), IoT, industrial internet of things, cloud computing, cognitive computing, artificial intelligence and machine learning [3,4].

Industry 4.0 addresses the transition from implanted frameworks to digital actual frameworks. In Industry 4.0, implanted frameworks, the internet of things (IoT), machine-to-machine communication and CPS technologies are incorporated into virtual reality with the reality. Furthermore, there is another stage of modern frameworks that includes keen plants that can manage the intricacy of creation in a digital actual climate. Exploration delineates how Industry 4.0 addresses the methodology of the fourth Industrial Revolution, whereupon ICT structures the infrastructural establishment for the upcoming imaginative mechanical innovations. In this new ICT-driven innovative development, implanted frameworks, IoT, CPS, industrial integration, and industrial information are the most significant components. Moreover, there are various other technologies that can be used for implementing Industry 4.0. These technologies include CPS, IoT, cloud computing, blockchain, industrial information integration, and other related technologies [2,5].

Industry 4.0 offers the manufacturing firms new methods to examine and designs to use; manufacturing, procurement, and inventory data that will help them accomplish their vision of modernization and efficiency. The schematics of the evolution of industry 4.0 are depicted in Fig. 11.1. Industry 4.0 utilizes psychological registering procedures alongside industrial internet of things (IIoT) applications. It applies to information science and insightful models to investigate continuous information from different machines, cycles, and frameworks; and afterward, it robotizes fabricating as needs to be. Different assembling businesses have applied IoT and IIoT to propel creation, dispersion, transportation, administration, and upkeep in the assembling system. An advanced venture's activity includes various dynamic exercises, requiring a lot of data and serious calculation. At a certain point, fabricating ventures required various processing assets, for example, workers for data sets and dynamic units. This caused wasteful information trade and sharing, low usefulness, and less ideal usage of assembling assets. In such a scenario, distributed computing can be employed to address such issues [6–8].

A huge volume of information can be transferred to a distributed computing community for the capacity and calculation that works with assembling lines and production. In addition, cloud-based assembling is a developing innovation that has the potential to significantly contribute to the acceptance of Industry 4.0, which enables modularization and administration direction in the context of assembling, where framework coordination and sharing of administrations and parts are critical

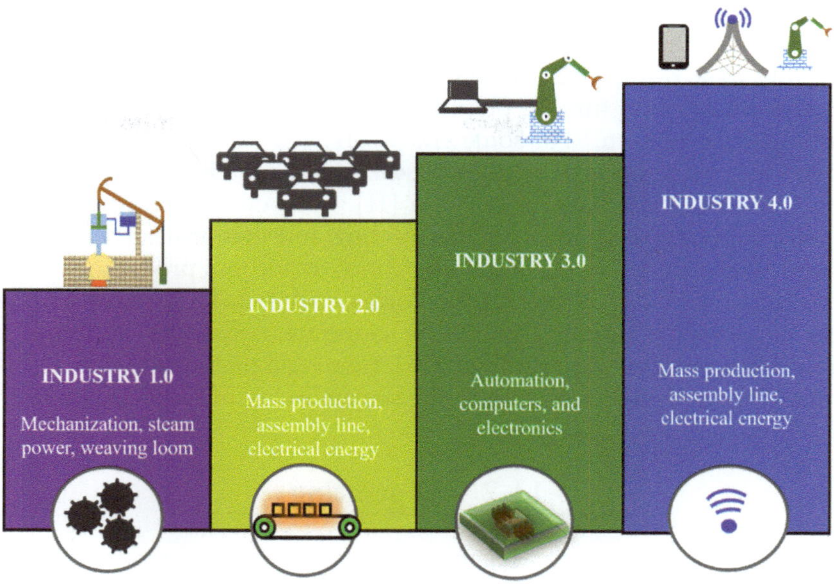

FIGURE 11.1 Evolution of Industrial Revolutions to Industry 4.0.

considerations. CPS is the center establishment of Industry 4.0. It is relied upon to give the premise to the production of Industrial IoT, which consolidates with cutting edge ICT to make Industry 4.0 feasible. CPS interfaces virtual space with actual reality by incorporating figuring, correspondence, and capacity abilities. Moreover, it tends to be constant, effective, solid, and secure [2,5,9,10].

Elements of the Industry 4.0 are even, vertical, and digital reconciliation of the whole framework. The key regions are standardization, administration of intricate frameworks, exhaustive foundation, security and privacy (information insurance), work association and plan (the jobs of the representatives are transformed, they are more Involved, progress is better and there is life-long learning), lawful structure (harmonization of legitimate frameworks), and the powerful utilization of assets (expected investment funds of crude materials and energy). Industry 4.0 is centered on making perceptive items, cycles, and techniques. In the brilliant assembly line, laborers, machines, and assets convey easily.

The internet of things and the internet of administrations, which refer to a global network of people, objects, and machines, are at the heart of corporate vision 4.0. After production, items, transportation hardware, and apparatuses "collaborate" to complete each phase. In this way, it could prompt the network of virtual worlds and actual items in the genuine world. Approaches to the trademark for Industry 4.0 are even integration, vertical incorporation, and coordinated advanced engineering [2,5].

Industry 4.0 promotes integrating cutting-edge ICT into manufacturing to improve productivity and capabilities. Interest in Industry 4.0 has risen recently as a result of the belief that the current state of Industry 4.0 represents a critical turning point in history. Industry 4.0 is at this point not a "future pattern." For some endeavors, it is presently at the core of their vital examination plan. Industry 4.0 has the capability of turning into the worldwide language of creation [11]. Each cycle utilized in an Industry 4.0 framework incorporates existing and demonstrated advancements with new advances and applications to address producing issues. The Industry 4.0 has different technologies, such as IoT, cloud computing, big data analytics, high processing computers, robotics/automation, skilled laborers, and 3D printing (3DP). Herein, in the present chapter, we present the role of 3D printing in the sustainability of Industry 4.0 [12].

11.2 3D printing

In the present era, many organizations face ever-increasing challenges in item improvement. The clients inquire for imaginative, attractive, and customized items of high quality at a sensible price. Similarly, the financial life expectancy of these goods decreases, which leads businesses to reduce their time to market and development cycles. Through globalization, the opposition in rich business sectors has increased. Imitators from foreign sectors make it harder for the organizations to keep up with market shares. One answer for increment advancement and an opportunity to advertise is conveyed by another production technique, additive manufacturing (AM). AM is currently transforming into a hub of innovation, with a growing consensus that 3D printing improvements will be one of the significant technical innovations [13].

3D printing is an AM technique for rapid prototyping of a wide range of monolithic structures (Fig. 11.2). The interaction consists of printing progressive layers of materials that are framed on top of one another. Three-dimensional printing, which encompasses a variety of methodologies, materials, and hardware, has progressed through time and has the potential to alter assembly and coordination procedures. Added substance production is widely used in a variety of industries, including development, prototyping, and biomechanical engineering [14].

11.2.1 Brief methodologies

Additive manufacturing techniques were developed to meet the necessity for printing complex patterns at fine goals. Quick prototyping assisted with the ability to print large structures, lowering printing

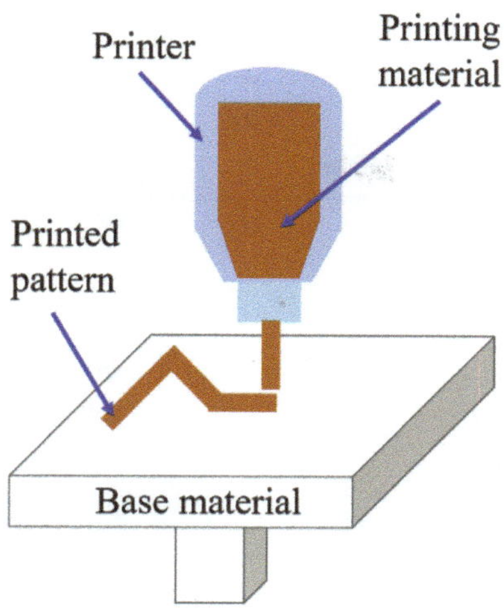

FIGURE 11.2 Schematics of a 3D printer and its components.

abandons, and improving mechanical characteristics is just a few of the important elements that have propelled AM forward. Fused Deposition Modeling (FDM) is one of the most popular 3D printing techniques that primarily makes use of polymer fibers. A persistent fiber of a thermoplastic polymer is used in the FDM process to 3D print layers of materials. The fiber is warmed at the nozzle to arrive at a semifluid state and afterward expelled on the stage or on top of recently printed layers. The thermoplasticity of the polymer fiber is a key feature of this method, as it allows the fibers to fuse during printing and then harden at ambient temperature thereafter. The thickness of the layer, width, and direction of fibers and air holes (in a similar layer or between layers) are the principal handling boundaries that influence the printed material's mechanical properties. The underlying cause of mechanical deficiency was shown to be interlayer distortion. Minimum expense, high velocity, and simplicity of cycle are the principal benefits of FDM. FDM, on the other hand, has a number of drawbacks, including brittle mechanical properties, poor surface quality, layer-by-layer appearance, and a limited selection of thermoplastic materials [15,16].

Powder bed fusion (PBF) measures are made up of thin layers of extremely fine powders that are distributed on a stage and forcefully pressed. A laser beam or a binder is used to mix the powders in each layer (Fig. 11.3). The next layers of powder are layered on top of the previous layers and mixed till the final 3D component is built. A vacuum is used to remove the excess powder, and if necessary, additional handling and specifications such as covering, sintering, or infiltration are carried out. The most critical factors for the feasibility of this approach are powder

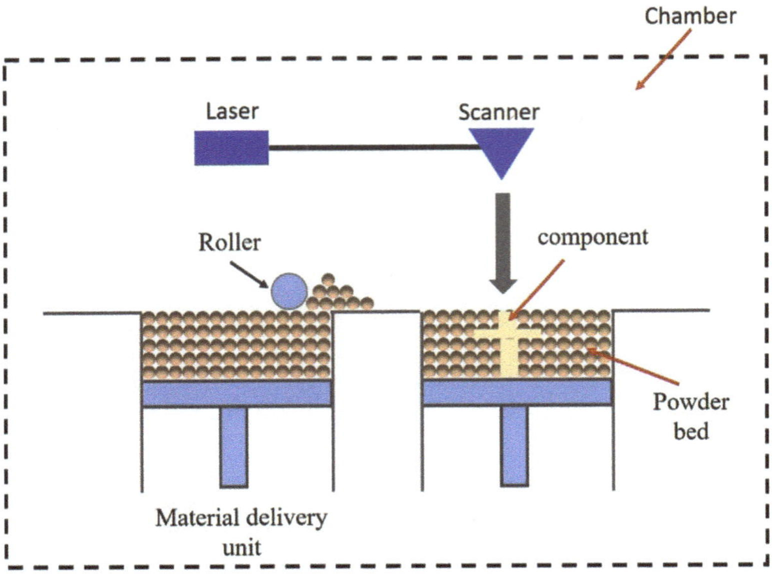

FIGURE 11.3 Schematics of powder bed fusion (PBF) 3D printing and its components.

size distribution and pressing that determine the thickness of the printed item. For powders with a low liquefying/sintering temperature, the laser should be used, but a fluid binder should always be used [17]. Selective laser sintering (SLS) may be used on a wide range of polymers, metals, and alloy powders, but selective laser melting (SLM) is limited to metals like steel and aluminum. The primary features of this approach are fine resolution and high-quality printing, which make it suitable for printing complex designs. This method is widely used in a variety of industries for cutting-edge applications such as tissue design frameworks, grids, airplanes, and gadgets. The main advantage of this method is that the powder bed is used as a support, which eliminates the need to remove supporting material [18,19]. In any situation, the main drawbacks of powder bed fusion, which is a slow process, include high costs and a high level of porosity when the powder is fused with a binder.

In the principal 3D printing procedure much of the time is utilized in photopolymerization, which overall alludes to the restoring of photoreactive polymers by utilizing a laser, light, or ultraviolet light. The case of 3D printing advances by utilizing photopolymerization and digital light processing (DLP). It uses UV light (or an electron beam) to initiate a chain reaction on a monomer layer. The monomers (mostly acrylic or epoxy-based) are UV-dynamic, and following radicalization, they immediately transform into polymer chains. Following polymerization, an example is cemented into the sap layer to hold the subsequent layers in place [20].

After the printing is finished, the unreacted resin is removed. For some printed items, a postmeasure treatment such as warming or photograph restoration may be used to provide the best mechanical exhibition. To print ceramic polymer composites or polymer-determined certifiable monomers, such as silicon oxycarbide, a scattering of clay particles in monomers can be used. SLA produces excellent products with a precise resolution of as little as 10 μm. However, it is somewhat sluggish, and expensive, and the range of materials available for printing is quite limited [21]. Furthermore, the response's energy and the relieving mechanism are complicated.

Direct Energy Deposition (DED) is a complex printing technique that is typically employed to repair or add material to existing components. It offers a high level of grain structure control and may provide excellent results. On a fundamental level, direct energy deposition is comparable to material ejection or powder bed combination (Fig. 11.4). The spout is not attached to a single pivot and can move in a variety of directions. Additionally, the interaction may be used with pottery. Polymers, on the other hand, are commonly used with metals and metal-based composites such as wire or powder. Laser deposition and laser engineered net shaping are two examples of this breakthrough (LENS). DED works by directing energy (laser or electron beam) into a small region of the substrate while simultaneously melting feedstock material (powder or wire) [22]. After passing the laser beam, the liquefied material is maintained and

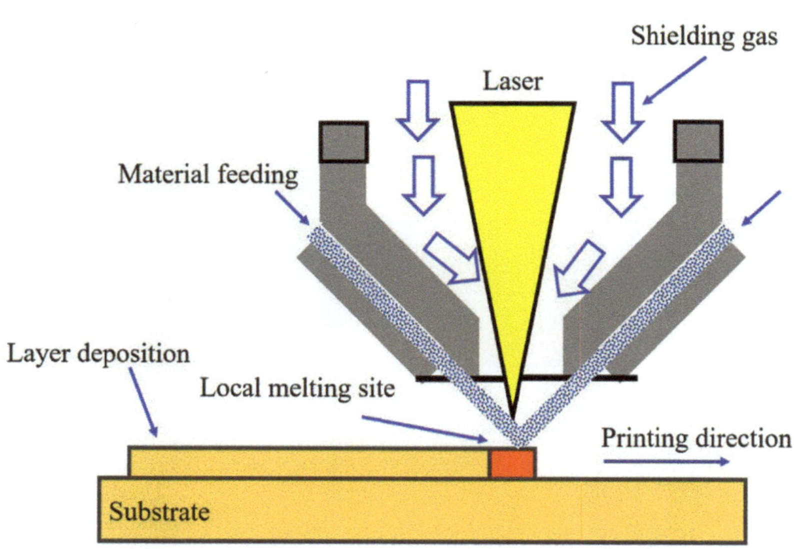

FIGURE 11.4 Schematics of Direct Energy Deposition (DED) 3D printing and its components.

intertwined into the dissolved substrate and set. The difference between DED and SLM is that DED does not use a powder bed, and the feedstock is liquefied before statement in a layer-by-layer approach similar to FDM, but with a significantly larger amount of energy for softening metals. As a result, it might be beneficial for filling breaks and retrofitting manufactured components when the use of the powder bed process is banned.

For aircraft applications, this method is often used with titanium, Inconel, hardened steel, aluminum, and comparable combinations. DED is characterized by high rates and extraordinarily large work envelopes in general. In comparison to SLS or SLM, it has a lesser precision (0.25 mm), a worse surface quality, and can create fewer complex objects. As a result, DED is frequently used for large components with minimal complexity, as well as for repairing larger parts. DED can reduce assembly time and cost while providing excellent mechanical qualities, a controlled microstructure, and precise arrangement control. This method may be used to repair turbine motors and other specialized applications in a variety of industries, including autos and aviation [22,23].

The Sheet Lamination process makes use of laminated object manufacture (LOM), which is one of the most cost-effective AM methods. It is dependent on layer-by-layer cutting and sheet overlay. Progressive layers are cut to perfection with a mechanical shaper or laser, and then reinforced (form-then-bond) or the other way around (bond-then-form) (Fig. 11.5). The form-then-bond technique is particularly useful for warm holding of earthenware and

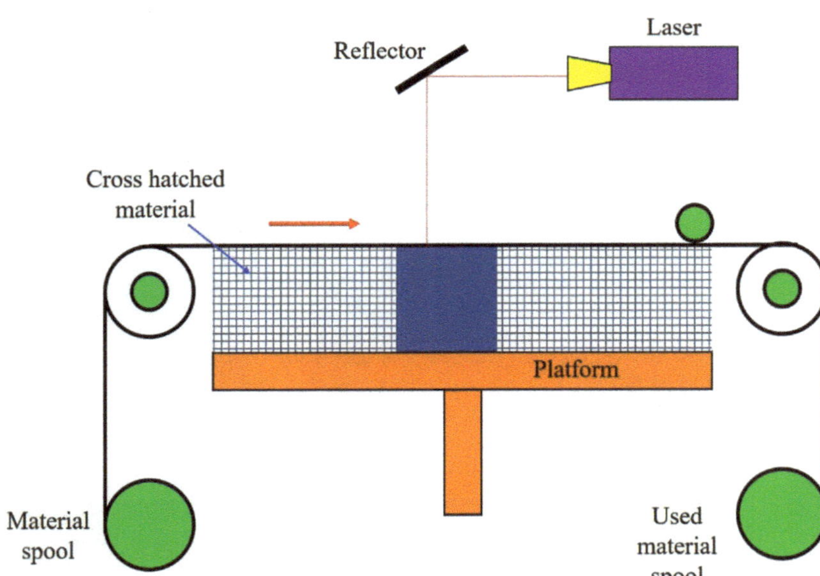

FIGURE 11.5 Schematics of Sheet Lamination 3D printing and its components.

metallic materials, and it also helps with the growth of interior components by removing excess materials before holding. Excess material is left in the aftermath of cutting for assistance, and once the cycle is completed, it can be removed and reused. Polymer composites, ceramics, paper, and metal-filled tapes are just a few of the materials that may be used with LOM. Post-processing, for example, high-temperature treatment might be required relying upon the sort of materials and wanted properties. LOM is one of the most amazing AM techniques for larger designs since it may reduce tooling costs and assembly time. Nonetheless, as compared to powder-bed fusion, LOM has average surface quality (without post-preparing) and worse dimensional accuracy. Similarly, when compared to powder bed procedures, removing the superfluous pieces of coverings after the object has been developed is time-consuming. As a result, it is not recommended for intricate forms [24–26].

Inkjet printing is one of the primary methods for additive manufacturing of ceramics. It is utilized for printing complex and progressed fired constructions for applications like platforms for tissue designing. In this strategy, a steady ceramic suspension, for example, zirconium oxide powder in water, is siphoned and kept as drops by means of the infusion spout onto the substrate. The droplets then form a structure that hardens to adequate strength to hold further resulting layers. This method is rapid and efficient, allowing for greater adjustability when developing and printing complicated structures. Wax-based inks and fluid suspensions are the two most common types of earthenware inks. Wax-based inks are dissolved and allowed to solidify on a cold substrate. Fluid suspensions, on the other hand, are toughened by evaporation. The nature of inkjet-printed components is determined by the molecular size dispersion of ceramics, the thickness of the ink and solid substance, expulsion rate, spout size, and printing speed. The major problems of this approach include the inability to keep up with functionality, the narrow objective, and the lack of connectivity between layers [27,28].

Contour crafting, called as comparable innovation to inkjet, is the principle strategy for AM of huge structures. Using larger spouts and strong pushing factors, this method is suitable for expelling large amounts of glue or dirt. It has been developed as a prototype for use on the moon. There are no debates over whether machine or innovation is more effective because each has its own set of uses [29]. Three-dimensional printing technology is no longer limited to prototype but is increasingly being used to create a wide range of things.

11.2.2 Materials for 3D printing

To construct trustworthy and exceptional gadgets, 3D printing requires top-notch materials that match reliable determinations. To ensure this, material controls procedures, requirements, and agreements are established between material sources, purchasers, and end-clients. Three-dimensional printing technology can generate fully functioning items out of a variety of materials, including ceramics, metals, polymers and their blends, composites, and functionally graded materials (FGMs) [30–32].

Because of the advantages provided by this connection, metal 3D printing innovation has grown in popularity in the aircraft, automobile, clinical application, and fabrication industries. Metals have excellent physical qualities, and they may be used in a variety of applications ranging from printing human organs to aviation parts. Aluminum composites, cobalt-based amalgams, nickel-based composites, tempered steels, and titanium amalgams are examples of these materials. In a 3D printed dental application, a cobalt-based composite is acceptable. This is due to its high explicit solidity, strength, high recovery limit, stretching, and heat-treated characteristics. Furthermore, 3D printing technology may be used to make aeroplane parts using nickel-base amalgams. Three-dimensional printed objects utilizing nickel base amalgams can be utilized in perilous conditions. This is due to its strong consumption resistance and the ability to conserve up to 1200°C in temperature. Finally, titanium mixtures may be used to create the product using 3D printing technology [14,23,32]. Titanium composites will have extremely specific qualities, such as pliability, high erosion resistance, oxidation resistance, and thinness. They are used in high-stress environments with high operating temperatures and anxiety, such as aviation parts and the biomedical sector.

Three-dimensional printing innovations are generally utilized for the creation of polymer parts from models to practical constructions. It can frame a 3D printed via the testimony of progressive layers of ejected thermoplastic fiber, for example, polylactic corrosive (PLA), acrylonitrile butadiene styrene (ABS), polypropylene (PP), or polyethylene (PE) using fused deposition modeling (FDM) (PE). In recent years, thermoplastic fibers with greater liquefying temperatures, such as PEEK and PMMA, have been as 3D printing materials. Because of their low cost, low weight, and versatility in handling, 3D printing polymer materials in a fluid condition or with a low dissolving point are widely used in the industry. Polymer materials, in general, played a vital role in biomaterials and clinical device items as idle materials, helping in the efficient operation of the devices as well as providing mechanical support in a variety of muscle enclosures.

Through the strengthening of the boundaries and organization of the outstanding mechanical characteristics, 3D printing technology can now construct patterned items using ceramics and cement without large pores or any fractures. Because of its liquid condition prior to setting, earthenware manufacture may be used in virtually any calculation and form, and is therefore ideal for future development and construction. According to them, ceramic materials are beneficial in dentistry and aviation applications. Alumina, bioactive glasses, and zirconia are examples of these materials. Three-dimensional Printing innovation, for example, may be able to quantify alumina powder. Alumina is a fantastic creative oxide with several applications, including impetus, adsorbents, microelectronics, synthetic substances, the avionics sector, and other high-innovation industries [33–35]. Alumina has incredible relieving intricacy.

Composite materials with unusual flexibility, low weight, and tailorable qualities have revolutionized the world of high-end manufacturing. Carbon strands made of polymer composites and glass strands made of polymer composites are examples of composite materials. Because of their high explicit solidity, strength, great consumption obstruction, and excellent exhaustion execution, carbon fiber built up polymers composite designs are widely used in the aeronautic trade. Glass filaments made of polymer composites, on the other hand, are widely used in 3D printing for a variety of purposes and have a wide range of prospective applications due to their cost-effectiveness and excellence. Fiberglass has a high coefficient of warm development and a strong warm conductivity. Furthermore, fiberglass is noncombustible and is unaffected by the high temperatures used in assembly procedures, making it an excellent option for use in 3D printing [24,26].

Keen materials are defined as materials that may modify the calculation and state of an object in response to external conditions such as heat and water. Self-advancing construction and sensitive mechanical technology framework are examples of 3D printed things made with sharp materials. Four-dimensional printing materials may also be outsourced to keen materials. Form memory chemicals and shape memory polymers are examples of gathering keen materials. Some shape memory combinations, such as nickel-titanium, can be used in biomedical analytics and in the development of small electromechanical devices. Changing temperatures, repeatability of microstructure, and thickness are key issues in the manufacture of 3D printed goods using nickel-titanium. Meanwhile, shape memory polymer (SMP) is a type of useful material that reacts to stimuli such as light, heat, a few types of substances, and so on. The jumbled condition of shape memory polymer might be useful in deliverer by applying 3D printing technology [36–38]. The dimensional exactness, surface roughness, and component thickness are used to determine the material's quality.

11.2.3 Role in present industries

Three-dimensional printing, as one of the techniques for fast prototyping, allows producers to create complicated items quickly and affordably. Models created with 3D technology are used in the inventive work stage to resolve design concerns prior to large-scale manufacture. It is a manufacturing cycle that is commonly used to depict the creation of more materials. Using a 3D printer, this cycle entails layering material, such as metal or plastic, to create a three-dimensional object. With the start of the 21st century, the world is seeing Industry 4.0 upheaval [3,10,13]. At the core of this chaos are shrewd assembling cycles, the digitization of industry, and capacities which are business-related. Three-dimensional printing is a critical part in the proliferation of the Industry 4.0 cycles.

However, since their introduction in the 1980s, 3D printers have been critical in the Industry 4.0 revolution. The rise of 3D technologies, or additive manufacturing, has been aided by the use of automated processes and technology in various sectors and industries. The use and execution of additive manufacturing, in combination with other innovations, is advancing the industry toward smart creation, in which machines (self-governing, programmed, and intelligent), frameworks, and organizations may exchange data and respond to the executives' creation frameworks. Furthermore, 3D printing plays a critical role as an invention capable of converting a 3D blueprint into an object without the use of intermediaries. Furthermore, the need for expensive instruments and apparatuses is eliminated, reducing post-handling, material waste, and human mediation. These characteristics describe the future's business. Industrial facilities may increase their flexibility thanks to 3D printing, allowing them to respond to the demands of an extremely demanding and erratic market. Furthermore, it enables the production of a wide range of bespoke products without the need of expensive forms and assembly equipment [35,37,38]. Furthermore, 3D printing is a fantastic climate partner, which is an important trademark given our current climatic situation and the need of having maintainable assembly methods that use less resources and generate less waste.

Three-dimensional printing has come to development rapidly as indicated by the notable Gartner curve. This curve addresses the development, reception, and business utilization of a particular innovation. At the most noteworthy pinnacle is clinical embedded 3D printing, retail location 3D printing and in supply chains. Likewise, we additionally discover 3D imprinting in study halls inside the schooling area and bioprinting for research in the field of medication. Overall, most manufacturing businesses will benefit from adopting 3D printing technology as quickly as possible. However, due to the significant expenditure reserve funds associated with digitizing their stock, the most evident advantages will be reaped by vehicle and large assembly companies. Customer producing, guard hardware manufacturers, and medical services organizations, notably those in dental medical services and prosthetics, are among the other businesses that are likely to gain first.

Finally, we are on the verge of a new mechanical revolution known as Industry 4.0. Three-dimensional printers are poised to play a key role in this advanced revolution of industry as their speed, reliability, health, and nature increase, and their cost drops. As the performance of 3D printers improves and their cost decreases, new freedoms will emerge, bringing 3D printing closer to large-scale production. As 3D printing advances, the range of products that may be created expands as well. The use of 3D printers in Industry 4.0 will be aided by the rapid improvement of particular printing materials, the combination of computerized security to safeguard IP, and the certification of 3D things by administrative organizations. The capacity of creative manufacturers who choose to adopt the principles of Industry 4.0 and digitalize their organizations quickly, on the other hand, will undoubtedly gain the most. It is also used in a number of areas such as the transportation industry, food industry, health and safety sector, etc. As the manufacturing takes place layer-by-layer, it is very easier to form complex designs which is not possible with the traditional methods. Three-dimensional printing involves the use of plastic majorly, which is one of the lighter materials than other metals. This benefits the aerospace and automotive sector where lightweight products are needed, which will eventually enhance the fuel efficiency. It is also used widely in the medical sector and electronics sector. Research and Development in the medical sector involves production of new machines and devices that can be made through 3D printing. Transportation sector is rapidly adapting this method as it is cost-effective and major parts which are made up of plastic are easily manufactured.

The use of 3D printing for film partition, desalination, and water purification applications has exploded in recent years, potentially changing the profession. This audit focuses on current advancements in 3D-printed materials and tactics for water-related applications, such as module spacers, innovative filtration and desalination layers, adsorbents, water remediation, solar orientated steam age materials, catalysis, and so on. The introduction of new 3D printers with higher printing goals, more competency, faster speeds, and wider material compatibility has piqued interest and may change inventive work in this industry.

Three-dimensional printing takes into account more proficient vehicle model plan, prototyping, testing, and creation utilizing mechanical 3D printing programming. This product empowers fashioners to make printable plans which are the main critical stage in making 3D printed vehicle parts, and the center of the printing system. Three-dimensional printing overseen by a product that answers your own particular requirements instead of the normal of the entire business can prompt more massive expense investment funds (faculty costs, gathering costs, stock expenses, and so forth) and a significant reduction in underway time.

Three-dimensional printing programming can robotize the creation cycle and guarantee lean assembling.

For vehicles, trains, transports, 3D printing could permit customization. It may very well be an extraordinary technique to fabricate vehicles adjusted to their current circumstance or to the travelers, with exceptional supplies. Added substance assembling could likewise be a decent method to modify and upgrade plans to make lighter vehicles, for instance. Also, 3D printing is an ideal technique to supplant spare parts. When dealing with an old vehicle, tracking down the right extra part can be a bad dream or just excessively costly. Because of 3D printing, it is feasible to make spare parts uniquely designed without any problem. That is precisely how Jaguar dealt with recreating the exemplary XKSS model from 1957.

11.2.4 Advantages and limitations

Manufacturers are increasingly using 3D printing, often known as additive manufacturing. Because of some of the revolutionary benefits it may give, demand is increasing. This creation interaction offers a scope of benefits contrasted with conventional assembling techniques. These benefits incorporate those identified with configuration, time, and cost, among others. It considers the layout and printing of more complex designs than standard assembling procedures. Traditional cycles have plan constraints, which do not affect the use of 3D printing at present time. It can also generate parts quickly, which speeds up the prototyping process. This takes into account each stage in order to complete them more quickly. When compared to machining models, 3D printing is more cost-effective and faster at creating components since the part may be created in hours, taking into consideration each design modification to be completed at a significantly faster pace. Another advantage of print on demand is that it does not need a lot of storage space, unlike traditional assembly methods. This saves space and money since there is no compelling need to print in bulk unless it is absolutely necessary. Because the 3D plan papers are produced using a 3D model as either a CAD or STL file, they are completely stored in a virtual library and can be searched and printed whenever needed. Changes to plans may be done at a very minimal cost by changing individual papers instead of wasting obsolete material and spending money on instruments.

Plastic is the most common 3D printing medium, while several metals can also be used. Plastics, on the other hand, have the advantage of being lighter than their metal counterparts. This is especially important in industries like automobiles and aircraft, where lightweightness is a concern and can transmit more ecofriendliness. Parts can also be created from custom-made materials to provide certain features, such as heat resistance, increased strength, or water repellency.

Depending on the design and complexity of a segment, 3D printing may manufacture products in minutes, which is far faster than formed or machined components. Furthermore, by producing STL or CAD documents suitable for printing, the plan interaction may be completed in a very short time. When compared to elective techniques, which are cut from huge chunks of nonrecyclable materials, the manufacture of components requires just the resources necessary for the actual part, with virtually no waste. Not only does the cycle save money on assets, but it also lowers the costs of the materials that are used.

Three-dimensional printing, as a single advanced assembly method, reduces time and hence costs associated with using several equipment. Three-dimensional printers may also be set up and left to continue working on their own, eliminating the need for administrators to be present at all times. As previously stated, this assembly approach may also save money on resources because it only uses the amount of material required for the actual item, with virtually no waste. While purchasing 3D printing gear might be pricey, you can save this charge by transferring your project to a 3D printing administration company.

Three-dimensional printers are turning out to be increasingly more available with more nearby specialist co-ops offering rethinking administrations for assembling work. This saves time and doesn't need costly vehicle costs contrasted with more conventional assembling measures created abroad in nations like China. This relationship is fundamentally innocuous to the ecology since it reduces the amount of material waste used. However, when you add aspects like improved ecofriendliness from using lightweight 3D printed parts, the environmental benefits are multiplied.

Furthermore, 3D printing is being used in the healthcare field to help save lives by producing human organs such as livers, kidneys, and hearts. Further advances and uses are being made in the medical services field giving probably the greatest advances from utilizing the innovation. As every technology also has some challenges to face and some issues to resolve, 3D printing also has some limitations. Notwithstanding the advantages of 3D printing like the opportunity of plan, customization, and the capacity to print complicated constructions, there are a couple of disadvantages that would need more research and innovative. These disadvantages incorporate significant expenses, restricted applications in enormous constructions and large-scale manufacturing, second rate and anisotropic mechanical properties, restriction of materials and deformities.

There are few challenges that the 3D printed items have to undergo. One of them is the administrative-related issues. All biomedical items require FDA endorsement. The biomedical sector is currently concentrating on Class I devices, which require less resources to sustain. Nonetheless, progress on Class II and III devices is continuing, with the approval of certain Class II inserts. The next issue is the limited number of materials that may be used. Biomaterials that are often used in 3D printing cannot be 3D printed, and the top performing AM materials are not biocompatible. As a result, there has been a huge increase in the development of new tactics and materials. Furthermore, 3D printed objects for specific uses may have inconsistencies in product quality. AM materials' mechanical characteristics have not been adequately characterized. AM materials and interaction boundaries have a huge impact on the final characteristics.

Alongside, void arrangement is one of the fundamental downsides of 3D printing that could be observed between resulting layers of materials. Because of the decreased interfacial holding between printed layers, the additional porosity created by AM might be exceedingly high, reducing mechanical execution. The degree to which voids are arranged is very dependent on the 3D printing approach and the printed material. The arrangement of voids is more regular and regarded one of the key surrenders that results in second rate and anisotropic mechanical qualities in processes that use fibers of materials, such as FDM or form forming. Delamination between layers can also occur as a result of this void arrangement after printing [15,29].

One of the most difficult aspects of AM is dealing with anisotropic microstructures and mechanical characteristics. In light of the concept of layer-by-layer printing, the microstructure of the material within each layer differs significantly from that at the layer's boundaries. Under vertical pressure or pressure assessed to that level, anisotropic conduct results in the mechanical conduct of the 3D printed item being unique. The expansion of the subsequent layers heats the limits of the previous layers in metals and amalgams 3D printed by heat combination (SLS or SLM), resulting in an alternative grain microstructure and anisotropic conduct due to heated angles. The laser pillar's warm penetration into each layer is a key aspect in managing the sintering system and limiting anisotropic conduct.

11.2.5 Technological advancements

The improvement of programmed strategies to join CT-filter results and configuration examinations with AM innovations will consider producing patient-explicit embeds rapidly. Furthermore, the development of drugs and medication delivery systems will be based on the patient's needs and characteristics. A biomedical embed's key components include complicated pieces with desirable mechanical qualities, cell connection and development, conveyance of supplements, waste, and antitoxins, biocompatibility, and bio-resorb ability. With new strategies, AM might improve both of these features at the same time. Metallic inserts coated with ceramic coatings, for example, can be made with cutting-edge functionality and expertise. While many new efforts are being done in the field of bioprinting and in-situ printing, many more are being done to upscale bioprinted frameworks and tissues for clinical applications and to investigate the cost viability of AM for tissue design. In the future, it may be possible to repair organs and tissues in situ. The research into putting together AM artificial organs, calculating vascularization, innervation, and accomplishing the multi-usefulness provided by each organ is ongoing. These organs are likely to be combined with electronic devices (cyborg organs), such as the bionic ear, which was used as an inductive curl to obtain electromagnetic signals for hearing.

The significant challenges in assembling speed, precision, repeatability, and cost, related to 3D printing, make it unsuitable for large-scale manufacturing in many industrial plants. Nonetheless, in the manufacture of sophisticated and customized products, it outperforms traditional manufacturing processes. Decentralization may be achieved in the future by dispersing workload among plants/machines through the use of cloud services, which will also aid in increasing production speed. Because the manufacturing industry will be distributed to many different sites such as tiny workplaces or houses, the roles of architects, processing facilities, and clients will be dramatically redefined. In other words, personal and personalized manufacturing will overcome the existing impediment of mass production on location.

The mechanical weakness of 3D printed items is discovered to be due to improper bonding between the different printed layers that make up the 3D part. Plastics are often utilized for extrusion 3D printing in fused deposition modeling. This method involves squeezing material from a nozzle, which prints pieces layer by layer, cools, and then joins them to make the finished 3D item. According to studies, these layers contain join imperfections, and printed products appear to be weaker than injection-molded parts. Additional heating is necessary to connect these interfaces more fully. Only the interfaces between printed layers were to be heated, not the entire object, to prevent the 3D printed part from warping and melting and losing its form. Carbon nanotubes were used to increase interlayer bonding. As the carbon particles heated in reaction to electrical currents, these nanomaterials were deposited on the surface of each printed layer. Currents must overcome a short gap of air between the print head and the 3D component in order to apply electricity when the thing is being produced. Metal electrodes that directly touch the printed component are one method for bridging the air gap; however, this might cause accidental damage to the part. As a result, a stream of charged air particles, or plasma, was created that could transfer an electrical charge to the printed part's surface. Electric currents were able to travel through the printed object, heating the nanotubes and welding the layers together using this method. Localized heating was used to strengthen 3D-printed parts, allowing them to match the mechanical qualities of molded ones.

Residual stress, which occurs as a result of the repetitive heating and cooling of metal printing processes, can cause faults in components and, in rare situations, printer damage. The impacts of varied printing patterns on titanium alloy parts created using a standard laser-based process were investigated to better understand how residual stress accumulates and how it might be reduced. The study focused on laser powder bed fusion (LPBF), a common additive manufacturing technology in which a laser scans over a layer of metal powder in a preset pattern, melting and fusing particles at the surface together. As the molten metal solidifies, a support stage lowers, and the printer applies a new coat of powder on top, letting the laser to continue layer by layer constructing the item. Residual tension begins to appear once the second layer of a structure is completed. Because the metals used in LPBF cool fast, the metal from the previous layer is solid by the time the printer's laser starts heating up a new layer. As the molten layers cool, they shrink inward, drawing on the solid metal underneath and causing stress. The bigger the temperature difference, the more the melted layer pulls. This procedure is repeated for each layer until the item is finished, resulting in the stresses being locked into solid metal. A continuous scan is the simplest basic printing pattern in LPBF, in which the laser scans back and forth from one end of the component to the other. The theory behind this method is that melting little portions of metal, or islands, one at a time rather than a full layer at a time will result in less metal contracting at the same time, lowering overall stress.

11.3 3D printing and Industry 4.0—potential and future scope

Business 4.0 has drawn the attention of both academics and industry in the previous decade, since it is seen as a major paradigm change in future manufacturing. AM, as a vital technology in the context of the impending revolution, holds a lot of promise for future advances in this new age if some existing limitations can be overcome in the near future. Industry 4.0 is steadily gaining traction, influencing manufacturing, employment, and a variety of other facets of life. Integrating such aspects into teaching or education is critical because it may help graduates develop the skills needed for Industry 4.0. Decentralization may become possible in the future by transferring responsibility for manufacturing plants/machines through the effective usage of cloud administrations. Another prospective aspect of AM is the manageability issue, in which AM might play a key role in minimizing waste assets and energy consumption by adopting just-in-time creation. In addition, 3D printing and digital assembling may expectedly affect the general public. To begin with, the job of representative in the business is to be reclassified to such an extent that they perform occupations about administration/plan/investigation as opposed to being in the workforce. Second, the stages like do-it-yourself and creator development work with clients are to be included in the plan and assembling stage. For example, students can plan their own items by transforming the study hall into a little active lab with a modest 3D printer. New material compounds for enhanced microstructures, imaginative design systems for appropriate boundary assessment, improved Computer Aided Design utilities for enhancement/reproduction/displaying purposes, novel AM/crossbreed measures along with continuous measure control and review, and so on are a few well-known research fields in additive manufacturing in the era of Industry 4.0. The primary suggestion is for the exploration community, industry, and legislators to work together to overcome the present obstacles on AM. Furthermore, AM's in effect new invention raises concerns of field normalization, necessitating additional fitting confirmation work. Nontraditional assembly techniques should have been developed due to the demand for mass customization in Industry 4.0. In this way, AM might turn into a critical innovation for manufacturing modified items because of its capacity to make modern articles with cutting edge credits (new materials, shapes). AM is presently being used in a variety of industries, including aviation, biomedical, and manufacturing, because of its high product quality. Despite certain concerns regarding its materiality in large-scale manufacturing, the usage of additive manufacturing (AM) in the industry is on the rise as a result of new mechanical advancements. It may create a way of supplanting typical manufacturing methods sooner rather than later, since it is a developing innovation for making accurate and strengthened complicated goods with speed.

11.4 Conclusions

The ability to configure, mass-customize, and create complex structures with little waste are three of 3D printing's main advantages. A comprehensive review of 3D printing strategies, materials, and current progress in moving applications in various businesses was done. The fundamental challenges associated with the concept of 3D printing have been overly discussed. As far as strategies, melded statement demonstrating (FDM) is quite possibly the most well-known. It was originally designed for 3D printing polymer fibers, but it has since been modified to work with a variety of different materials. When compared to powder-bed techniques like specific laser sintering and specific laser dissolving, FDM is primarily used for rapid prototyping, and the mechanical qualities and quality of the printed items are inferior (SLM). Powder bed methods use a helper cement to mix, soften, or strengthen neighboring powders, resulting in better results but higher costs and slower cycles. Direct energy testimony (DED) uses a source of energy (laser or electron bar) to dissolve metal powders; however unlike SLM, there is no powder bed, and the feedstock is softened before statement in a layer-by-layer way similar to FDM, but with a far larger amount of energy for softening metals. Inkjet printing is extremely quick and is used for 3D printing of fired suspensions, although it necessitates the use of heat medications thereafter. Forming, which is based on the ejection of materials (concrete), is used to print larger structures such as buildings. Stereolithography is one of the pioneering 3D printing processes, mostly for photopolymers that can provide components with a very fine resolution. In any event, it is a slow and complicated technique with a finite amount of ingredients. Finally, layer-by-layer cutting and overlay of sheets or material transfers are required for overlaid object fabrication (LOM). Fibers, wire, powder, glue, sheets, and inks are examples of materials that may be used in 3D printing. Polymers are the most well-known materials that have been developed for rapid prototyping. The most often used polymers for 3D printing are acrylonitrile-butadiene-styrene (ABS) copolymers, polyamide (PA), polycarbonate (PC), and polylactic corrosive (PLA), thermosetting powders such as polystyrene, and polyamides and photopolymer tars. The addition of stranded polymers and nanomaterials improved the mechanical characteristics of the 3D printed composite, allowing it to be used as a practical material. Metals are essentially powders (or wires), and the major 3D printing processes are SLS, SLM, and DED. Due to a limited selection of metals and combinations suitable for 3D printing, demands for the application of existing processes to a broader range of amalgams and composite designs have been made. Earthenware manufacture has developed a pattern for tailoring materials with a high solidarity-to-weight ratio and has experimented with the creation of delicate clay cross-sections for applications such as tissue design burned frameworks. The basic difficulty, however, is a limited amount of materials that can be used to 3D print pottery with better control over the microstructure and organization of the item. Moreover, despite the fact that the development industry has been hesitant to adopt 3D printing of cement, mass customization, no necessary formwork, and computerization are all promising features. A substantial combination with great stream capacity, measure capacity, mechanical execution, and appearance has been the fundamental space of late turns of events. Added substance fabricating (AM) generously added to the new exploration and advancement of biomaterials for prototyping complex and altered constructions with patient-explicit necessities. In any case, it confronts challenges such as limited resources and administrative constraints. The avionics industry has invested in additive manufacturing to promote customized components with improved solidity-to-weight ratios, as well as quick plane support and on-demand fabrication. Furthermore, AM's acceptance in the aerospace industry confronts challenges like limited resources and high costs, as well as the contradictory nature of 3D printed parts. AM innovation in the development sector is still in its early phases, with a limited number of productive tasks throughout the world. When compared to traditional tactics, the main disadvantages are high costs and poor mechanical execution. Despite the benefits of additional material fabrication, there are a few drawbacks that would necessitate further imaginative effort to implement this technology in other endeavors. Due to a decrease in interfacial holding between printed layers, void growth between subsequent layers of materials causes additional porosity during the manufacturing process, which might reduce mechanical execution. Another common AM difficulty is anisotropic conduct, which results in different mechanical conduct under vertical strain or pressure as compared to the even heading. Moving CAD into a 3D-printed item, on the other hand, frequently results in inaccuracies and deformities, particularly in bended surfaces, owing to CAD's decorating notion, which is an estimate of the design. Furthermore, AM's layer-by-layer look in applications such as constructions, toys, and aviation isn't ideal. Progressing innovative work of materials and techniques have served to bypass a portion of these difficulties; however, there is still opportunity to get better. Despite being a forward-thinking technique for customized items and specialized applications, 3D printing requires further innovation to compete with traditional methods in large-scale manufacture of standard products due to its higher cost and slower pace. In any event, AM has advanced tremendously in recent years. The increased funding and inventive work from around the world will hasten the transition from traditional assembly tactics to 3D printing.

References

[1] L. Da Xu, E.L. Xu, L. Li, Industry 4.0: state of the art and future trends, Int. J. Prod. Res. 56 (8) (2018) 2941–2962, https://doi.org/10.1080/00207543.2018.1444806.

[2] L.S. Dalenogare, G.B. Benitez, N.F. Ayala, A.G. Frank, The expected contribution of Industry 4.0 technologies for industrial performance, Int. J. Prod. Econ. 204 (August) (2018) 383–394, https://doi.org/10.1016/j.ijpe.2018.08.019.

[3] E. Oztemel, S. Gursev, Literature review of Industry 4.0 and related technologies, J. Intell. Manuf. 31 (1) (2020) 127–182, https://doi.org/10.1007/s10845-018-1433-8.
[4] A.C. Pereira, F. Romero, A review of the meanings and the implications of the Industry 4.0 concept, Procedia Manuf. 13 (2017) 1206–1214, https://doi.org/10.1016/j.promfg.2017.09.032.
[5] A.G. Frank, L.S. Dalenogare, N.F. Ayala, Industry 4.0 technologies: implementation patterns in manufacturing companies, Int. J. Prod. Econ. 210 (January) (2019) 15–26, https://doi.org/10.1016/j.ijpe.2019.01.004.
[6] S. Vaidya, P. Ambad, S. Bhosle, Industry 4.0 - a glimpse, Procedia Manuf. 20 (2018) 233–238, https://doi.org/10.1016/j.promfg.2018.02.034.
[7] C.S. Tang, L.P. Veelenturf, The strategic role of logistics in the industry 4.0 era, Transp. Res. Part E Log. Transp. Rev. 129 (June) (2019) 1–11, https://doi.org/10.1016/j.tre.2019.06.004.
[8] M. Javaid, A. Haleem, R. Vaishya, S. Bahl, R. Suman, A. Vaish, Industry 4.0 technologies and their applications in fighting COVID-19 pandemic, Diabetes Metabol. Syndr. Clin. Res. Rev. 14 (4) (2020) 419–422, https://doi.org/10.1016/j.dsx.2020.04.032.
[9] M. Ghobakhloo, Industry 4.0, digitization, and opportunities for sustainability, J. Clean. Prod. 252 (2020) 119869, https://doi.org/10.1016/j.jclepro.2019.119869.
[10] U.M. Dilberoglu, B. Gharehpapagh, U. Yaman, M. Dolen, The role of additive manufacturing in the era of industry 4.0, Procedia Manuf. 11 (June) (2017) 545–554, https://doi.org/10.1016/j.promfg.2017.07.148.
[11] S. Khan, Fossil Fuel and the Environment, IntechOpen, 2012.
[12] C. Bai, P. Dallasega, G. Orzes, J. Sarkis, Industry 4.0 technologies assessment: a sustainability perspective, Int. J. Prod. Econ. 229 (2020) 107776, https://doi.org/10.1016/j.ijpe.2020.107776.
[13] T.D. Ngo, A. Kashani, G. Imbalzano, K.T.Q. Nguyen, D. Hui, Additive manufacturing (3D printing): a review of materials, methods, applications and challenges, Compos. B Eng. 143 (December 2017) (2018) 172–196, https://doi.org/10.1016/j.compositesb.2018.02.012.
[14] Q. Yan, et al., A review of 3D printing technology for medical applications, Engineering 4 (5) (2018) 729–742, https://doi.org/10.1016/j.eng.2018.07.021.
[15] P. Tack, J. Victor, P. Gemmel, L. Annemans, 3D-printing techniques in a medical setting: a systematic literature review, BioMed. Eng. Online 15 (1) (2016) 1–21, https://doi.org/10.1186/s12938-016-0236-4.
[16] X. Wang, M. Jiang, Z. Zhou, J. Gou, D. Hui, 3D printing of polymer matrix composites: a review and prospective, Compos. B Eng. 110 (2017) 442–458, https://doi.org/10.1016/j.compositesb.2016.11.034.
[17] I. Gibson, D. Rosen, B. Stucker, Additive Manufacturing Technologies: 3D Printing, Rapid Prototyping, and Direct Digital Manufacturing, second ed., 2015.
[18] S. Sun, M. Brandt, M. Easton, Powder Bed Fusion Processes: An Overview, Elsevier Ltd, 2017.
[19] S. Vock, B. Klöden, A. Kirchner, T. Weißgärber, B. Kieback, Powders for powder bed fusion: a review, Prog. Addit. Manuf. 4 (4) (2019) 383–397, https://doi.org/10.1007/s40964-019-00078-6.
[20] A. Bagheri, J. Jin, Photopolymerization in 3D printing, ACS Appl. Polym. Mater. 1 (4) (2019) 593–611, https://doi.org/10.1021/acsapm.8b00165.
[21] M. Layani, X. Wang, S. Magdassi, Novel materials for 3D printing by photopolymerization, Adv. Mater. 30 (41) (2018) 1–7, https://doi.org/10.1002/adma.201706344.
[22] W.J. Oh, W.J. Lee, M.S. Kim, J.B. Jeon, D.S. Shim, Repairing additive-manufactured 316L stainless steel using direct energy deposition, Opt. Laser Technol. 117 (February) (2019) 6–17, https://doi.org/10.1016/j.optlastec.2019.04.012.

[23] J.D. Majumdar, D. Madapana, I. Manna, 3-D printing by laser-assisted direct energy deposition (LDED): the present status, Trans. INAE 6 (4) (2021) 933–953, https://doi.org/10.1007/s41403-021-00252-9.

[24] S.C. Ligon, R. Liska, J. Stampfl, M. Gurr, R. Mülhaupt, Polymers for 3D printing and customized additive manufacturing, Chem. Rev. 117 (15) (2017) 10212–10290, https://doi.org/10.1021/acs.chemrev.7b00074.

[25] M.R. Khosravani, T. Reinicke, On the environmental impacts of 3D printing technology, Appl. Mater. Today 20 (2020), https://doi.org/10.1016/j.apmt.2020.100689.

[26] T. Singh, S. Kumar, S. Sehgal, 3D printing of engineering materials: a state of the art review, Mater. Today Proc. 28 (xxxx) (2020) 1927–1931, https://doi.org/10.1016/j.matpr.2020.05.334.

[27] S. Patra, V. Young, A review of 3D printing techniques and the future in biofabrication of bioprinted tissue, Cell Biochem. Biophys. 74 (2) (2016) 93–98, https://doi.org/10.1007/s12013-016-0730-0.

[28] K. Tappa, U. Jammalamadaka, Novel biomaterials used in medical 3D printing techniques, J. Funct. Biomater. 9 (1) (2018), https://doi.org/10.3390/jfb9010017.

[29] R. Walczak, K. Adamski, Inkjet 3D printing of microfluidic structures - on the selection of the printer towards printing your own microfluidic chips, J. Micromech. Microeng. 25 (8) (2015), https://doi.org/10.1088/0960-1317/25/8/085013.

[30] I. Karakurt, L. Lin, 3D printing technologies: techniques, materials, and post-processing, Curr. Opin. Chem. Eng. 28 (2020) 134–143, https://doi.org/10.1016/j.coche.2020.04.001.

[31] I. Blanco, The use of composite materials in 3d printing, J. Compos. Sci. 4 (2) (2020), https://doi.org/10.3390/jcs4020042.

[32] J.W. Stansbury, M.J. Idacavage, 3D printing with polymers: challenges among expanding options and opportunities, Dent. Mater. 32 (1) (2016) 54–64, https://doi.org/10.1016/j.dental.2015.09.018.

[33] Z. Chen, et al., 3D printing of ceramics: a review, J. Eur. Ceram. Soc. 39 (4) (2019) 661–687, https://doi.org/10.1016/j.jeurceramsoc.2018.11.013.

[34] L.C. Hwa, S. Rajoo, A.M. Noor, N. Ahmad, M.B. Uday, Recent advances in 3D printing of porous ceramics: a review, Curr. Opin. Solid State Mater. Sci. 21 (6) (2017) 323–347, https://doi.org/10.1016/j.cossms.2017.08.002.

[35] T. Wang, X. Lu, A. Wang, A review: 3D printing of microwave absorption ceramics, Int. J. Appl. Ceram. Technol. 17 (6) (2020) 2477–2491, https://doi.org/10.1111/ijac.13604.

[36] M. Shunmugasundaram, M.A. Ali baig, M. Ajay Kumar, A review of bio-degradable materials for fused deposition modeling machine, Mater. Today Proc. 27 (xxxx) (2020) 1596–1600, https://doi.org/10.1016/j.matpr.2020.03.267.

[37] J. Saroia, et al., A review on 3D printed matrix polymer composites: its potential and future challenges, Int. J. Adv. Manuf. Technol. 106 (5–6) (2020) 1695–1721, https://doi.org/10.1007/s00170-019-04534-z.

[38] K.A. Eley, Centralised 3D printing in the NHS: a radiological review, Clin. Radiol. 72 (4) (2017) 269–275, https://doi.org/10.1016/j.crad.2016.12.013.

Index

Note: Page numbers followed by "*f*" indicate figures and "*t*"indicate tables.

D

E

N

O

X

Z

FSC
www.fsc.org
MIX
Papier | Fördert
gute Waldnutzung
FSC® C083411

Kolibri 360 GmbH
Ferdinand-Jühlke-Straße 7
99095 Erfurt, Deutschland
produktsicherheit@kolibri360.de